Cities and Nature

Cities and Nature fosters high-quality multi-disciplinary research addressing the interface between cities and the natural environment. It provides a valuable source of relevant knowledge for researchers, planners and policy-makers. The series welcomes empirically based, cutting-edge and theoretical research in urban geography, urban planning, environmental planning, urban ecology, regional science and economics. It publishes peer-reviewed edited and authored volumes on topics dealing with the urban and the environment nexus, including: spatial dynamics of urban built areas, urban and peri-urban agriculture, urban greening and green infrastructure, environmental planning, urban forests, urban ecology, regional dynamics and landscape fragmentation, among others.

More information about this series at http://www.springer.com/series/10068

Karsten Grunewald · Junxiang Li
Gaodi Xie · Lennart Kümper-Schlake
Editors

Towards Green Cities

Urban Biodiversity and Ecosystem Services in China and Germany

Springer

Editors
Karsten Grunewald
Leibniz Institute of Ecological Urban and Regional Development
Dresden
Germany

Gaodi Xie
Institute of Geographic Sciences and Natural Resources Research
Chinese Academy of Sciences
Beijing
China

Junxiang Li
School of Ecological and Environmental Sciences
East China Normal University
Shanghai
China

Lennart Kümper-Schlake
Bundesamt für Naturschutz (BfN)
Bonn, Nordrhein-Westfalen
Germany

ISSN 2520-8306 ISSN 2520-8314 (electronic)
Cities and Nature
ISBN 978-3-319-86353-5 ISBN 978-3-319-58223-8 (eBook)
DOI 10.1007/978-3-319-58223-8

Printed on acid-free paper

This Springer imprint is published by Springer Nature
The registered company is Springer International Publishing AG
The registered company address is: Gewerbestrasse 11, 6330 Cham, Switzerland

Foreword

Today's urbanization processes cause losses of open land at the margins of urban agglomeration and compaction processes in city centers. These developments lead to the transformation and often a destruction of ecosystems, and result in the loss of flora and fauna from previously biodiverse habitats. At the same time, humans' well-being in cities is progressively being challenged by increasing environmental pollution and the risks of negative climate-induced impacts.

Under these circumstances, urban green spaces can offer part of the solution. They can contribute to improve air quality, reduce heat in summer, buffer negative effects of heavy precipitation events, and serve as sites for recreation. But urban green spaces can also provide habitats for numerous species. In fact, cities can act as biodiversity hotspots. Thus, they allow for nature experience and environmental education right at our doorsteps, and thereby increase the acceptance for conservation efforts elsewhere. Eventually, nature-based solutions may solve challenges urban planners and local governments are confronted with.

The need for conservation and restoration of urban ecosystems, such as urban forests and wetlands, urban parks or temporary brownfields is increasingly being acknowledged by international initiatives and captured in related targets. This refers, for example, to the 2030 Agenda for Sustainable Development (SDGs) or the United Nations Habitat III Conference on Housing and Sustainable Development. In the Strategic Plan for Biodiversity 2011–2020 of the Convention on Biological Diversity (CBD), a number of targets are related to the manifold facets of urban biodiversity conservation and ecosystem restoration. Besides, as reflected in Chapter 2 in this book, national strategies and initiatives like the Chinese New-Type Urbanization Plan or the Urban Green process in Germany are developing. They highlight the relevance of nature-based solutions to address urban challenges relevant to growing cities in China, Germany, Europe, and beyond.

Since nature-based solutions represent a rather novel approach to sustainable urban development, this book aims to capture the significance and values of urban biodiversity and ecosystem services and to describe and convey them also to an audience who is not familiar with the benefits and potential of urban green (Chap. 3).

The implementation of different approaches seems to be hampered by a lack of knowledge and information. Chapter 4, therefore, explores options and barriers for a successful integration of various approaches into planning processes and instruments for enhancing green spaces under land scarcity. Illustrative examples from different Chinese and German cities demonstrate how to strategically plan, finance, and successfully increase urban green space.

Recommendations extracted from this book's case studies and an analysis of further challenges for the development of green cities reveal great potential for further cooperation between China and Germany in order to mutually foster innovations for nature-based solutions (Chap. 5).

Innovations are needed in this rather new field of applied sustainability research. For a long time, urban development policy and sciences have neglected the potential of urban green spaces, and nature conservation policy and sciences also only considered the topic at the margins. This has changed in recent years. Accordingly, we are glad to be in the position to generate new impulses for urban planning, design, and development.

First ideas for a green cities study emerged in 2015 at the 8th Sino-German Workshop on Biodiversity and Nature Conservation, jointly organized by the Chinese Research Academy of Environmental Sciences (CRAES) and the German Federal Agency for Nature Conservation (BfN). First results were presented at the 5th Sino-German Environmental Forum held in 2016 in Nanjing, organized by the Chinese Ministry for Environmental Protection (MEP) and the German Ministry for the Environment, Nature Conservation, Building and Nuclear Safety (BMUB). We appreciate contributions from different scientific disciplines and feedback from agencies of various ministries and nonacademic organizations to several drafts. The joint generation of knowledge as well as inter- and transdisciplinary modes of work are crucial for fostering innovation and providing tangible recommendations for urban planners, decision-makers, and stakeholders on how to increase and enhance urban green spaces in quantity and quality.

We hope that this book may serve as an impulse from the view point of nature conservation to various stakeholders, including the Sino-German Partnerships on Urbanization and the Environment. We thank the authors, especially from IOER, ECNU, CUMT, University Salzburg, and IGSNRR/CAS, for their contributions and the Springer publishing house for including this book in their new Series "Cities and Nature".

Beijing, China Prof. Li Haisheng, CRAES
Bonn, Germany Prof. Beate Jessel, BfN

Acknowledgements

This book is an outcome of the project "Towards Green Cities: The Values of Urban Biodiversity and Ecosystem Services in China and Germany". The project has been initiated and commissioned by the German Federal Agency for Nature Conservation (BfN) and is implemented by a research consortium comprising the East China Normal University (ECNU), the University of Salzburg, the China University of Mining and Technology (CUMT) and has been led by the Leibniz Institute of Ecological Urban and Regional Development (IOER). The project has been supported by the Institute of Geographic Sciences and Natural Resources Research, Chinese Academy of Sciences (IGSNRR, CAS) and the Chinese Research Academy of Environmental Sciences (CRAES). This study contributes to the Sino-German Urbanization Partnership.

We would kindly like to thank all the authors for providing English text of contributions and Dr. Felix Pahl, Berlin for the language polishing. We thank especially Sabine Witschas, Dipl.-Ing. Cartography at IOER for improving the figures of the book, Burkhard Schweppe-Kraft and Beyhan Ekinci at BfN for reviewing subchapters, and Natalja Leutert at IOER for checking the references. The organization and the cooperative effort, with Springer as editor, were carried out in a notably pleasant atmosphere.

Key Issues

- *'Towards green cities' means ecologically sound development*, a crucial issue in general and within the international competition of cities for sustainable urban development. But many challenges remain when it comes to implementation. Is urban green space development consistent with a compact city which is argued to be a sustainable urban form for reducing urban sprawl and the consumption of land resources within the city and its fringes? Is it possible to reconcile modern urban living, a growing infrastructure, attractive work, and recreational opportunities with preservation of biodiversity and a balanced supply of green spaces? (Chaps. 1 and 5).
- *'Green' urban development figures prominently on the political agenda in China and Germany*. We review recent challenges and concepts of urban development at the global, national, and municipal level, assuming that the integration of urban biodiversity and urban ecosystems into urban planning strategies is essential for human well-being in general. However, we understand that all planning activities, also concerning the implementation of urban green infrastructure should be particularly tailored to the specific characteristics of a city. Additionally, we present in which context current approaches of the two countries can be compared (Chap. 2).
- 'Green matters'—we not only celebrate our appreciation for nature, we demonstrate the *multiple benefits as well as nonmonetary and monetary values of urban green space* for health and well-being, environmental justice and recreation as well as for the regulation of climate, water, and air in cities. Valuation of ecosystem services can highlight the cost-effectiveness and profitability of investments in nature-based solutions. Additional economic arguments enable a new promising perspective and contribute to better consideration of environmental aspects in decision making. How 'green' are our cities and how much 'green space' is enough? The establishment of target values is difficult. In addition to the proportion and volume of green space, spatial arrangement, accessibility, and other, especially qualitative aspects play a key role (Chap. 3).

- The evidence- and indicator-based assessment of urban biodiversity enables the *integration of ecosystem service values into policy and planning*. We show common features and differences between the Chinese and German planning and financing system for urban green space as well as options, forms, and perceptions for the green and blue infrastructure elements for different purposes. What can we learn from each other regarding green city development? In which areas both countries can be forerunners? (Chap. 4).
- *Case study examples from China and Germany* with facts, lessons, successful projects, new trends of green spaces, and biodiversity being integrated into the city development show the contribution of nature-based solutions to air quality improvement, water regulation, soil protection, or climate change adaptation and mitigation and often provide cost-effective alternatives to technological options (Chaps. 3 and 4).
- Forward-looking *city planning strategies should safeguard urban green space in quantity and quality* to provide biodiversity and ecosystem services for the well-being of the population even in growing cities. For the overall understanding it is important to accept the coexistence of green and gray infrastructure; for the planning side, this requires integrated approaches, regulations and tools, action plans, and funding available; and to address implementation challenges, participation of local initiatives, overall communication and cooperation are prerequisites. Addressing urbanization processes with nature-based solutions is a key challenge of the twenty-first century in China, Germany, and elsewhere. This will require bringing all allies on board (Chap. 5).

Contents

Contributors

Martina Artmann Leibniz Institute of Ecological Urban and Regional Development, Dresden, Germany, e-mail: m.artmann@ioer.de

David Baier Amt für Stadtgrün, City of Bonn, Germany, e-mail: david.baier@-bonn.de

Olaf Bastian Environmental Office, Dresden, Germany, e-mail: olaf.bastian@web.de

Stephan Becsei B-A-E-R Urban + Environmental Research, Frankfurt am, Germany, e-mail: mail@b-a-e-r.com

Jürgen Breuste University Salzburg, Salzburg, Austria and East China Normal; University Shanghai, Shanghai, China, e-mail: juergen.breuste@sbg.ac.at

Jiang Chang China University of Mining and Technology, Xuzhou, China, e-mail: changjiang102@163.com

Boping Chen World Future Council, Beijing, China, e-mail: boping.chen@-worldfuturecouncil.org

Nannan Dong Tongji University, Shanghai, China, e-mail: dongnannan@tongji.edu.cn

Shanshan Feng China University of Mining and Technology, Xuzhou, China, e-mail: fengshanshan@cumt.edu.cn

Karsten Grunewald Leibniz Institute of Ecological Urban and Regional Development, Dresden, Germany, e-mail: k.grunewald@ioer.de

Bernd Hansjürgens Helm holtz Centre for Environmental Research–UFZ, Leipzig, Germany, e-mail: bernd.hansjuergens@ufz.de

Adrian Hoppenstedt HHP Hannover, Hannover, Germany, e-mail: hoppenst-edt@hhp-raumentwicklung.de

Wei Hou Chinese Academy of Surveying and Mapping, Beijing, China, e-mail: houwei@casm.ac.cn

Tinghao Hu China University of Mining and Technology, Xuzhou, China, e-mail: hth11211@163.com

Dennis Kalisch Technische Universität Berlin, Berlin, Germany, e-mail: d.kalisch @campus.tu-berlin.de

Lennart Kümper-Schlake Federal Agency for Nature Conservation (BfN), Bonn, Germany, e-mail: lennart.kuemper-schlake@bfn.de

Junxiang Li East China Normal University and Shanghai Key Laboratory of Urbanization and Eco-Restoration, Shanghai, China, e-mail: jxli@des.ecnu.edu.cn

Xiushan Li Chinese Research Academy of Environmental Sciences, 100012 Beijing, China, e-mail: xiushanli@vip.163.com

Song Liu Tongji University, Shanghai, China, e-mail: liusong5@tongji.edu.cn

Yuelai Liu Tongji University, Shanghai, China, e-mail: liuyuelai@gmail.com

Pingjia Luo China University of Mining and Technology, Xuzhou, China, e-mail: luopingjia@cumt.edu.cn

Juliane Mathey Leibniz Institute of Ecological Urban and Regional Development, Dresden, Germany, e-mail: j.mathey@ioer.de

Jonas Michels Amt für Stadtgrün, Bonn, Germany, e-mail: jonas.michels@bonn.de

Stefanie Rößler Leibniz Institute of Ecological Urban and Regional Development, Dresden, Germany, e-mail: s.roessler@ioer.de

Alice Schröder Federal Agency for Nature Conservation (BfN), Leipzig, Germany, e-mail: alice.schroeder@bfn.de

Anne Seiwert Leibniz Institute of Ecological Urban and Regional Development, Dresden, Germany, e-mail: a.seiwert@ioer.de

Xiaoliang Shi Shenyang Agricultural University, Shenyang, China, e-mail: sxl42 2127@163.com

Ralf-Uwe Syrbe Leibniz Institute of Ecological Urban and Regional Development, Dresden, Germany, e-mail: r.syrbe@ioer.de

Wolfgang Wende Leibniz Institute of Ecological Urban and Regional Development, Technische Universität Dresden, Dresden, Germany, e-mail: w.wende @ioer.de

Henry Wüstemann Technische Universität Berlin, Berlin, Germany , e-mail: henry.wuestemann@tu-berlin.de

Nengwen Xiao Chinese Research Academy of Environment Science, Beijing, China, e-mail: xiaonw@craes.org.cn

Suili Xiao Foreign Economic Cooperation Office, Ministry of Environmental Protection of the People's Republic of China, Beijing, China, e-mail: xiaosuili @163.com

Gaodi Xie Institute of Geographic Sciences and Natural Resources Research, Chinese Academy of Sciences (CAS), Beijing, China, e-mail: xiegd@igsnrr.ac.cn

Qiaoqiao Xu East Asia Secretariat ICLEI, Seoul, South Korea, e-mail: qiaoqiao. xu@iclei.org

Biao Zhang Institute of Geographic Sciences and Natural Resources Research, Chinese Academy of Sciences (CAS), Beijing, China, e-mail: zhangbiao@igsnrr.ac. cn

Yi Zhang Institute of Urban Environment, Chinese Academy of Sciences (CAS), Xiamen, China, e-mail: yizhang2015@hotmail.com

Hongxuan Zhou China University of Mining and Technology, Xuzhou, China, e-mail: zhouhongxuan@live.cn

Chapter 1
Introduction to an Urban Ecosystem Approach

Karsten Grunewald, Junxiang Li, Gaodi Xie and Lennart Kümper-Schlake

In the context of ongoing urbanization processes and sustainable urban development, the aim is to capture, describe, and convey to various target groups the current significance, the values, and the potentials of urban biodiversity and ecosystem services. To seek sustainable pathways, the current developments and different approaches are to be studied globally, whereas this book focuses on current processes and practices in China and Germany in more detail. The strategic goal is a long-term appreciation of the potentials and increased consideration of urban green spaces as nature-based solutions in city planning and development. What are predominant processes and most pressing issues? What can we learn from each other regarding 'green city development'? What are central terms in this field and how are they defined in this book?

K. Grunewald (✉)
Leibniz Institute of Ecological Urban and Regional Development, Dresden, Germany
e-mail: k.grunewald@ioer.de

J. Li
East China Normal University and Shanghai Key Laboratory of Urbanization and Eco-Restoration, Shanghai, China
e-mail: jxli@des.ecnu.edu.cn

G. Xie
Institute of Geographic Sciences and Natural Resources Research, Chinese Academy of Sciences (CAS), Beijing, China
e-mail: xiegd@igsnrr.ac.cn

L. Kümper-Schlake
Federal Agency for Nature Conservation (BfN), Bonn, Germany
e-mail: lennart.kuemper-schlake@bfn.de

K. Grunewald et al. (eds.), *Towards Green Cities*, Cities and Nature,
DOI 10.1007/978-3-319-58223-8_1

1.1 Background and Objectives

Lennart Kümper-Schlake, Boping Chen and Karsten Grunewald

Urbanization is a major syndrome of global change, and at the same time a key driver of global environmental change. Besides demographic, economic, and societal shifts, urbanization processes have major implications for ecosystems in cities, their hinterlands, and remote areas (MEA 2005; UNDP 2012). For decades, professionals in environmental sciences and nature conservation policies have focused on the protection and management of large-scale ecosystems and decision-makers managing urbanization processes have neglected the relevance of urban green spaces (Elmqvist et al. 2013). However, due to increased environmental problems and risks linked to converted ecosystems, in recent years this has been changing. Not only environmental scientists, urban designers and planners but also decision-makers for policies and businesses started considering the relevance and potentials of urban ecosystems for both integrated nature conservation and supporting public well-being (Wu 2014). This volume therefore presents recent developments to further support mainstreaming and implementation efforts, as well as protection and restoration activities of urban ecosystems with a specific focus on China and Germany.

Urbanization Processes and Changing Urban Ecosystems

Since 2008, more than half of the world's population has been living in cities, and by 2050 two-thirds of humankind is expected to live in agglomerations, which amounts to 6.3 billion urban dwellers, and it is expected that more than 60% of the urban area that these people will live in is yet to be built (UN 2014a, b). To date, the urban share of the population in many developed countries, such as Germany, is around 75%, but Asia, and in particular China, has the highest rate of change (Table 1.1). China's urbanization level has increased from 35.9% in 2000 to 49.2% in 2010 (UN 2014a, b), and it is expected to exceed three quarters of the overall population in 2050. After some years of less dynamic migration, German major cities are witnessing growth and re-densification processes again (Sect. 2.3). Globally, these migration processes combined with changing household and

Table 1.1 Global urbanization—status and prediction (*Source* UN 2014a, b)

Region/Country	Proportion urban (%)			Average annual rate of change (%)
	1990	2014	2050	2010–2015
Asia	32	48	64	1.5
China	**26**	**54**	**76**	**2.4**
Europe	68	69	78	0.1
Germany	**73**	**75**	**83**	**0.3**
Northern America	75	81	87	0.2
Latin America	71	80	86	0.3
Africa	31	40	56	1.1

consumption patterns lead to increased land take for settlement and transport in the urban fringes and to concentration and compaction processes in the city centers—including in China and Germany (WBGU 2016).

While cities only occupy about 2% of the earth's surface, the activities and needs of the urban population are responsible for 60–80% of the global energy consumption, 75% of carbon emissions, 60% of residential water consumption and the use of three quarters of timber (Grimm et al. 2008; SDG 2015). The urban ecosystems and the natural and near-natural surroundings of the population centers are being subjected to sometimes massive pressure through changed land-use and emissions into air, water and soil, leading to impairment of the ecosystem functions and corresponding services. These developments lead to risks for the functioning of —especially mega cities'—economies and societies (Kraas 2008). These risks are relevant for the local level, but also for a global level, with a view to the safe operating space for human development and well-being (WBGU 2016).

Seto and Reenberg (2013) explore five major trends in the urbanization process that are likely to have implications for biodiversity and ecosystem services, but at the same time might offer opportunities for implementing ecologically valuable green spaces in cities (Box 1.1). Many of these global observations also hold true for China and Germany and will be further explored in this book.

Box 1.1 Urbanization trends and impacts (cf. Seto and Reenberg 2013):

(1) The physical extents of urban areas are expanding faster than urban populations. This trend is based on the assumption that more land needs to be built up and urban populations will continue to increase. At some point, this is no longer true, since urban areas are also shrinking and are in need of restructuring and urban renewal (WBGU 2016).
(2) Urban areas modify their local and regional climate through the urban heat island effect and by altering precipitation patterns.
(3) Expansion of built-up areas will draw heavily on natural resources, in particular water (such as break of river connections), timber and energy. The land-use change at urban fringes often consumes agricultural land, with knock-on effects on habitats, biodiversity, and ecosystem services elsewhere.
(4) Urban land expansion is occurring fast in areas adjacent to biodiversity hotspots and faster in low-elevation, biodiversity-rich coastal zones than in other areas.
(5) Most future urban expansion will occur in areas of limited economic development and institutional capacity, which will constrain abilities to invest in the protection of biodiversity and the conservation and restoration of ecosystems.

Urban Biodiversity and Ecosystem Services

As cities continue to sprawl and change in size and shape, science and policy are recognizing the need to also protect, restore, and design urban ecosystems. This understanding is mainly based on the potentials urban green infrastructure can offer for increasing the cities' resilience towards climate change impacts, for well-being and quality of life as well as for the conservation and management of urban nature/biodiversity.

Contrary to common perceptions, cities are biodiversity hotspots due to their diverse structure and micro-climates (Kowarik 2005). In general, all forms of urban biodiversity can be an object of nature conservation and management. This may vary from natural remnants and traditional cultural landscapes to designed parks and gardens and urban-industrial nature (Sect. 3.2). While natural and traditional cultural landscapes are legally protected in Germany and China, and through ecological red-lining also in China, newly developed and designed parks offer great potentials to better integrate recreational functions and habitat conservation, especially given that structural diversity is likely to increase the attractiveness for recreational use (Sects. 4.2 and 4.3). Urban-industrial nature has been shown to be highly diverse, but the public's recognition towards it remains ambivalent (BMUB 2015). Regardless of their uniqueness, these urban spaces remain under great pressure from building activities in the process of re-densification (Schröder et al. 2016).

Besides the focus on urban biodiversity, the international scientific community started to apply the human-centered approach of ecosystem services on a broad scale over a decade ago (MEA 2005). It aims at highlighting the multiple benefits humans obtain from intact ecosystems, at the same time also highlighting the value and cost-effectiveness compared to conventional technical solutions. With the support of international and national initiatives like TEEB (The Economics of Ecosystems and Biodiversity; TEEB 2010; TEEB DE 2016), the policy arena is now starting to make use of first results and supports further research in related fields.

One major field that urban ecosystem services can be linked to is climate change adaptation. Ecosystem-based approaches are being used ever more frequently to increase the resilience of the urban fabric. Urban flood plains or lowered parks can serve as recreational areas, where urban dwellers spend their leisure time, but at the same time, these green spaces can help to buffer flood events or storm waters, leave space for air circulation and reduce urban heat island effects (Sect. 3.3). The underlying intention to promote nature-based solutions is based on the insight that old-fashioned planning, including soil sealing and ecosystem destruction, has increased the cities' vulnerability to risks. Illustrative case studies from Chinese and German cities demonstrate these facts (Chaps. 3 and 4).

Well-being and quality of life are increasingly being acknowledged in political and scientific discussions on urban green infrastructure. Human well-being increasingly is becoming a primary focus of urban sustainability projects (Bai et al. 2014), and urban green spaces play a major role in terms of their recreational function and an increase in the quality of life. Urban parks, for example, are highly welcomed by urban dwellers and are being used to spend their leisure time (BMUB 2015). Besides personal health benefits in terms of well-being, reduction of stress and decrease of high blood pressure, urban green is also a factor in improving the attractiveness of cities that compete for specialists and young talent (Sects. 3.1 and 4.2).

As cities are social-ecological systems with the highest density of human population that interact with various ecosystem types, ranging from natural remnants to novel urban ecosystems in relatively small areas, they are crucial for environmental education, for knowledge transfer on ecological processes and compartments. Thus, urban ecosystems increase the acceptance for nature conservation not only in cities but also elsewhere (Schröder et al. 2016).

International Cooperation: Partnerships on Sustainable Urbanization

When it comes to exchange of scientific and applied knowledge on urban ecology and the implementation of urban green spaces, international collaboration is beneficial for increasing the speed and quality of decisions and municipal measures being taken. Best practices and reporting of difficulties during the implementation process offer valuable insights for adjusting and improving one's own initiatives, plans and programs. Based on these premises, the China–ASEAN partnership for ecologically friendly development (and jointly pursued green development) and the EU-China Urbanization Partnership have been initiated. In addition to these partnerships, China's Ministry of Housing and Urban-Rural Development (MoHURD) and Germany's Federal Ministry for the Environment, Nature Conservation, Building and Nuclear Safety (BMUB) have signed the Sino-German Urbanization Partnership to deepen bilateral exchange also in the fields of urban ecosystem protection, rehabilitation and restoration (Sect. 5.3). The German Federal Agency for Nature Conservation (BfN) supports this endeavor together with partnering institutions in Germany, China and beyond.

These partnerships contribute to the needs of European and Asian experts, policymakers and city planners to translate political goals of sustainable urban development into practical actions (Sects. 5.1 and 5.2). The potential for international and inter- as well as transdisciplinary cooperation in creating 'eco-friendly cities' is immense. Also, China's leading international advisory body on questions of sustainable development, CCICED (China Council for International Cooperation on Environment and Development), has addressed questions on ecosystems and their management (Chen et al. 2014) as well as good city models under the concept of ecological civilization (CCICED 2014) in the past—but did not yet consider urban development and ecosystem management together.

Aim and Structure of the Book

This book aims at highlighting relevance, values, and potentials of urban biodiversity and ecosystem services for different target groups (policy-makers, urban planners, academics, public and private decision-makers, etc.) against the backdrop of sustainable urban development in ongoing urbanization processes. Recent developments that support mainstreaming and implementation of urban ecosystems as well as different approaches in China and Germany are discussed, to finally allow long-term appreciation of the potentials and increased consideration of urban green spaces in city planning and development. Figure 1.1 gives an overview of the structure of the study and related research and application issues.

Urbanization as a driver means not only an increase of population in cities, but also a growing demand of dwellers for housing space, infrastructure, and comfort as well as mobility, leisure activities, and others. What is the role of urban nature in this context? Which services and values does green space provide and does that

1 Background, Objectives, Key Terms of the Study

overall driver and questions

2 Challenges, Concepts and Strategies, City Characteristics

How is the topic of urban biodiversity and ecosystem services embedded in international, European and national (Chinese and German) regulations, processes, initiatives and practices of urban development and nature conservation?
Is it possible to compare Chinese and German cities?

3 Values and Benefits of Urban Biodiversity and Ecosystem Services

Which services of urban ecosystems are important for the quality of life in cities and relevant to policies and society in Germany and China?
Which evidence-based examples can be identified with respect to distribution, significance and values of urban biodiversity and urban green?

4 Implementing Green Spaces in Cities (Plans, Instruments, Options)

Which options can be pointed out and recommended for including green spaces in urban planning and development while taking into account established urban open space systems, target values, planning instruments, procedures and current constraints on implementing measures for enhancing green spaces – despite increasing scarcity of land?
What can we learn from examples of good practice?

5 Recommendations for Green Cities Development

and challenges ahead

Fig. 1.1 Scheme of the structure of the book 'Towards Green Cities' and selected research questions. © Karsten Grunewald

correspond with the demand and needs of dwellers? How must political steering at the national and municipal level be oriented, and which recommendations does science have to offer? How 'green' are our cities, and how 'green' should they be? Guided by these questions, the Chinese-German author team of this exploratory study tries to provide evidence-based arguments for practitioners in favor of green city development.

1.2 Key Terms of the Book

Compiled by Martina Artmann, Olaf Bastian, Jürgen Breuste, Karsten Grunewald, Juliane Mathey, Stefanie Rößler, Alice Schröder, Anne Seiwert, Ralf-Uwe Syrbe and Qiaoqiao Xu and agreed by all co-authors of the book

Urban Area/Cities

According to the United Nations (UN), an **urban area** can be defined by at least one of the following attributes (UNICEF 2012):

- administrative or political boundaries (municipality)
- threshold population size (differs across the world, minimum for an urban settlement is typically around 2000 people)
- population density, economic function (e.g., a significant majority of the population is not primarily engaged in agriculture, or there is surplus employment)
- presence of urban characteristics (e.g., paved streets, electric lighting, sewerage)

Cities can be understood as permanent human settlements characterized by three dimensions and their interlinkages (Kuper and Kuper 1996; ARL 2005):

- City as economic and societal unit represents economic and social processes.
- City as political unit represents an administratively defined area.
- City as physical structure is defined by buildings, green spaces, infrastructure.

In the context of this book, we use the term 'cities' mainly to refer to administrative units.

Green Cities

'Green' is often used as a general term to describe environmental sustainability or eco-friendliness. Accordingly, the **green city concept** follows the vision and ideals of sustainable urban development, focusing on environmental issues to some extent. For instance, the 'Green City Index' can be measured (Siemens AG 2012).

Lindfield and Steinberg (2012) distinguish **green cities** (which means cities that have already achieved, or are moving toward long-term environmental sustainability in all of its aspects) from cities that continue to follow environmentally unsustainable development trajectories. Thus, 'Green' within the green city concept is not only defined by green spaces or vegetation elements such as trees or roadside greenery but also involves a range of sectors such as urban agriculture, waste water treatment, eco-innovations, sustainable traffic solutions, etc. (BMUB 2015), which are an integral part of sustainable urban development. Nevertheless, the green city concept comprises an understanding of a city where green spaces are a part of its infrastructure in their own right and therefore are naturally integrated into urban planning processes. For the link to the 'Eco-city' approach, see Sect. 2.2.

Following this understanding, green cities are an indispensable contribution to an ecological civilization, to national sustainable development. The term 'green city' is sometimes also used as a marketing term or label (e.g., 'European Green Capital Award' by the European Union).

In this book we use the term 'green city' to refer to a city which focuses on green spaces, nature, and biodiversity as crucial elements of a sustainable city.

Urban Nature, Urban Green, Urban Green Spaces

The terms 'urban nature', 'urban green' and also 'urban green space' are often used synonymously.

'Urban nature' is used to cover all aspects of the biotic environment, from individual species up to large green spaces (Schröder et al. 2016). Following the study TEEB DE, '**urban nature**' is defined as the entirety of all nature elements (lifeless elements, microbes, fungi, plants, animals) found in urban areas, including their functional relations (TEEB DE 2016). It comprises relics of natural and cultural landscapes as well as nature elements that were designed horticulturally or newly developed after profound changes of former site conditions, for example on urban-industrial brownfields.

The more narrow term '**urban green**' is often used to focus on urban green spaces used by human beings. Following the German 'Greenbook on Urban Green' (BMUB 2015), urban green includes all forms of green urban open spaces and vegetated buildings such as parks, cemeteries, allotments, brownfields, areas for sports and playing, street vegetation and street trees, vegetation around public buildings, areas of nature conservation, woodlands and forests, private gardens, urban agricultural areas, green roofs, and green walls as well as other open spaces.

Following this book, '**urban green spaces**' are usually understood as spaces, which are directly used for active or passive recreation, or indirectly used by virtue of their positive influence on the urban environment, are accessible to citizens and serve the diverse needs of citizens and thus enhance the quality of life in cities (URGE Team 2004; GreenKeys Team 2008). Urban green spaces are essential providers of ecosystem services and can be regarded as service providing units for urban residents (Wurster and Artmann 2014).

China still follows the regulation of 'Urban Green Areas' (CJJ/T 85-2002) issued by the Ministry of Housing and Urban-Rural Development of the people's Republic of China in 2002 (MoHURD 2002). The urban green areas refer to the natural surfaces and artificial green surfaces that include parks, plantation areas, buffer green areas, attached green space, and other forms of green space in cities.

Particularly in the German context, the term 'urban open spaces' is used. It describes areas which are predominantly 'open' or not built-up within the urban fabric. They can be relics of natural or agricultural landscapes or intended by design as vegetated or even paved areas, but free of buildings above ground (e.g., Richter 1981; Milchert 2003).

In Germany, a distinction is made between public and private green spaces according to ownership and usage which influence responsibility, management opportunities and accessibility. In contrast, China has a unique land-use system in which there are two types of land ownership: state-owned urban land and rural land owned by farmer collectives. People can only acquire the right of usage (Huang et al. 2017).

In this book, we will predominately use the term 'urban green spaces', referring to the entirety of biotic and nature elements in urban areas, including both public and private urban green spaces.

Green Infrastructure (GI): In the context of political objectives of the European Union, green infrastructure is defined as 'a strategically planned network of high-quality natural and semi-natural areas with other environmental features' (EC 2013; EEA 2014). According to the EU, this includes the green infrastructure in both rural and urban areas (ibid.). Nevertheless, there is an ongoing discussion whether the term should be more narrowly defined especially in terms of GI in the urban context. The green infrastructure approach serves *inter alia* to enhance green space networks to a coherent planning unit (Ahern 2007; Lennon 2015). In a broader understanding, green infrastructure describes the entirety of the open and green spaces as well as their benefits and services, whereas the benefits and services of green infrastructure within a narrower sense result from its network character. Essential features that most approaches to green infrastructure have in common are (e.g., Benedict and McMahon 2006; Pauleit et al. 2011; Rouse and Bunster-Ossa 2013; Davies et al. 2015) that they

- adopt a multi-scale and multi-object approach including all kinds of urban green at different scales,
- focus on connectivity between green spaces at different scales and aim for multi-functionality,
- integrate green and gray infrastructure.

Blue infrastructure, such as streams, lakes, ponds, artificial swales, and storm water retention ponds, is also part of the green infrastructure (EC 2011; Elmqvist et al. 2015).

Urban Biodiversity

Biodiversity is used as a generalized term to describe the diversity of ecosystems, populations, and species and the genetic diversity within them. According to the Convention on Biological Diversity (CBD) (UN 1992), the aim of nature conservation is to preserve this biodiversity worldwide. This also applies to cities/urban areas. In urban settings, however, there is a paradigm shift from more or less natural ecosystems and native species towards a wider focus on the entire range of urban ecosystems, their associated ecosystem services, and possible contributions to species conservation. Since biodiversity conservation is a worldwide concern, species and ecosystems are considered as especially valuable if they are not common, which applies to urban habitats as well.

In contrast to other ecosystems, in urban ecosystems human beings influence biodiversity more directly through habitat loss, habitat fragmentation, and the introduction of new species, and indirectly by changing urban climate, soils, hydrology, and biogeochemical cycles (Kowarik 2005). This makes urban biodiversity different from biodiversity outside of urban areas. Not only physical alterations to the urban environment directly affect biodiversity patterns, but to a very high degree and very specifically also socioeconomic activities. Nevertheless, also near-natural habitats can be found in urban areas (Sect. 3.2), which is of course part of urban biodiversity. All forms of urban biodiversity should be properly managed to let urban dwellers benefit from its ecosystem services. Urbanization as a major global trend raises the question to what extent species of animals and plants can survive in urban settings, how to conserve them, how to manage their habitats, and how to give people access to them and let them benefit without destroying them.

Urban Ecosystem Services (UES)

Ecosystem services (ES) are direct and indirect contributions of nature to human well-being (according to the Millennium Ecosystem Assessment, MEA 2005). ES can be categorized into provisioning services (e.g., food and raw material supply), regulating services (e.g., pollutant purification and erosion control) and cultural services (e.g., landscape aesthetics, recreation, and tourism). Supporting ES (e.g., soil formation or photosynthesis) make the services of the three other groups possible. We do not assess supporting ES separately because they are implicit in the delivery process of the other services.

Urban ecosystem services (UES) support the urban quality of life, including protection from natural hazards (TEEB 2010; Haase et al. 2014a, b). Cities are dependent on ecosystems beyond the city boundaries, but also benefit from internal urban ecosystems (Bolund and Hunhammar 1999). Urban ecosystems are mainly represented by different types of urban green spaces in the city. This includes in particular parks, urban forests, water bodies, cemeteries, vacant lots, gardens and yards, landfills, road trees, green roofs, and walls as well as valuable, mostly protected habitats, such as floodplain forests, fens, peat bogs, near-natural lakes/ponds, orchard meadows, herbaceous meadows in wet environments, etc.

Since the diversity of ecosystems and biotic associations is part of biodiversity, ES and biodiversity are often mentioned in the same breath (e.g., Ridder 2008; TEEB 2010). Biodiversity supports the 'functioning of ecosystems'; however, it can also be defined as an ES in its own right—the ecosystem service of providing biodiversity (Grunewald and Bastian 2015).

Societal value creation by the ecosystem is assessed by means of the ES concept, and evaluated in monetary and non-monetary terms. The monetary assessment of ES can promote commitment and raise awareness of decision-makers and planners for the preservation of nature by reflecting the economic value of a functioning ecosystem (Grunewald and Bastian 2015). Especially the demand side needs to be supported by socioeconomic data. However, to get a holistic picture of the ecosystem's value, non-monetary values must also be taken into consideration (Chap. 3). By making the multiple benefits of urban green or the impacts of ecosystem degradation visible, UES provide key links for integrating planning, management and governance practices that seek transitions to more sustainable cities, and play an important role for resilience in urban systems.

Urban Planning

Urban development comprises diverse processes of changes in the urban structure (ARL 2005). **Urban Planning** refers to the proactive steering of the spatial development of a city or urban area (cf. ARL 2005). This also includes the development of urban green spaces (Sect. 4.1).

Urban Green Space Planning as a sectoral discipline of spatial planning deals with the realization, financing, and design of green spaces in cities. Fields of action include the qualification of green spaces, environmental protection, and nature conservation. In addition to this, by facilitating and maintaining green open spaces it also fulfills a social commitment and contributes to the quality of life (Wenzel and Schöbel 2001).

Nature Conservation in Urban Areas

In Germany, nature conservation is defined in the German Federal Nature Conservation Act (§ 1 BNatSchG) as follows: Nature conservation serves for the protection of nature and landscape by reason of their own value and as a basis for human life and health, in settled as well as in unsettled areas. It refers to the safeguarding and development of biodiversity, of capacity and functionality of the natural balance and of its capacity for regeneration. Beauty and recreational values of nature and landscape are also included. Nature conservation in urban areas includes—besides the protection of species, ecosystems (biodiversity) and abiotic factors—societal, social, and cultural aspects as well (Schröder et al. 2016). This makes clear the contributions of nature conservation to recreation, to nature experience and human health (Rittel et al. 2014). Moreover, it is a central aspect in urban areas to inspire people for nature and its conservation by involving citizens actively, offering opportunities for nature experience as well as by communicating knowledge about nature and environment (Schröder et al. 2016). China's nature conservation activities in urban areas follow the China National Biodiversity Conservation Strategy and Action Plan (2011–2030) adopted in 2010 (Sect. 2.2.1).

The demand to combine the protection of biodiversity, natural resources, and functionality of the natural balance with the safeguarding and increase of human quality of life requires the development of green spaces that meet the different requirements (Hansen et al. 2012). Specific aims of urban nature conservation are: the provision of adequate opportunities for recreation and nature experience, observing wildlife, environmental education, research, the protection and development of open and green spaces as diverse habitats for flora and fauna, as well as the prevention of natural disasters.

Urban nature conservation has rapidly grown in importance over the past decades and will continue to do so in the coming years. In this context, cities should be regarded as a new type of environment with species compositions and habitats peculiar to urban-industrial areas. Especially in urban areas, nature conservation can be put into practice only if it is backed by a majority of the population or, at least, supported by a vociferous minority and tolerated benevolently by the silent majority (Wittig 1999).

Nature-based Solutions (NbS) are 'defined as actions to protect, sustainably manage, and restore natural or modified ecosystems that address societal challenges effectively and adaptively, simultaneously providing human well-being and biodiversity benefits.' (Cohen-Shacham et al. 2016). They are '(…) inspired by, supported by or copied from nature' (EC 2015) and support preserving and restoring ecosystems. NbS can be characterized as systemic solutions providing multidimensional benefits by simultaneously enhancing and securing human well-being in cities, promoting economic growth and protecting urban ecosystems. NbS overlap with other similar concepts (e.g., ecosystem services, green infrastructure, ecosystem-based adaptation) that all deal with analyses and responses that address pressures on and risks for urban and non-urban ecosystems. However, since NbS stand out for their focus on solutions for social challenges, they provide a valuable grounding for developing systemic solutions for the increasing complexity and social challenges especially cities are confronted with (e.g., climate change, resource consumption, or biodiversity loss).

References

Ahern J (2007) Green infrastructure for cities: the spatial dimension. In: Novotny V, Brown P (eds) Cities of the future. Towards integrated sustainable water and landscape management. IWA Publishing, London, pp 267–283

ARL—Akademie für Raumforschung und Landesplanung (2005) Handwörterbuch der Raumordnung. Akademie für Raumforschung und Landesplanung, Hanover

Bai X, Shi P, Liu Y (2014) Realizing China's urban dream. Nature 509:158–160

Benedict MA, McMahon ET (2006) Green infrastructure. Linking landscapes and communities. Island Press, Washington

BMUB—Bundesministerium für Umwelt, Naturschutz, Bau und Reaktorsicherheit (2015) Grün in der Stadt – Für eine lebenswerte Zukunft. Grünbuch Stadtgrün. Bundesministerium für Umwelt, Naturschutz, Bau und Reaktorsicherheit, Berlin

Bolund P, Hunhammar S (1999) Ecosystem services in urban areas. Ecol Econ 29:293–301

CCICED—Chinese Government, China Council for International Cooperation on Environment and Development (2014) Good city models under the concept of ecological civilization. CCICED Special Policy Study Report, Beijing

Chen Y, Jessel B, Fu B, Yu X, Pittock J (2014) Ecosystem services and management strategy in China. Springer

Cohen-Shacham E, Walters G, Janzen C, Maginnis S (eds) (2016) Nature-based Solutions to address global societal challenges. IUCN, Gland, Switzerland

Davies C, Hansen R, Rall E, Pauleit S, Lafortezza R, De Bellis Y, Santos A, Tosics I et al (2015) Green infrastructure planning and implementation. WP 5. Green Surge

EC—European Commission (2011) EU biodiversity strategy to 2020. European Commission, December 2011. http://register.consilium.europa.eu/doc/srv?l=EN&f=ST%2018862%202011%20INIT. Accessed 31 Aug 2015

EC—European Commission (2013) Building a green infrastructure for Europe. European Commission, Luxembourg

EC—European Commission (2015) Towards an EU Research and Innovation policy agenda for nature-based solutions & re-naturing cities. Final report of the Horizon 2020 expert group on 'nature-based solutions and re-naturing cities'. European Commission, Brussels

EEA—European Environment Agency (2014) Spatial analysis of green infrastructure in Europe. EEA Technical report, 2/2014. European Environment Agency, Copenhagen

Elmqvist T, Fragkias M, Goodness J, Güneralp B, Marcotullio PJ, Mc-Donald RI, Parnell S, Schewenius M, Sendstad M, Seto KC, Wilkinson C (ed) (2013) Global urbanisation, biodiversity and ecosystem services: challenges and opportunities. Springer, Berlin, Heidelberg, New York

Elmqvist T, Setälä H, Handel SN, van der Ploeg S, Aronson J, Blignaut JN, Gómez-Baggethun E, Nowak DJ, Kronenberg J, de Groot R (2015) Benefits of restoring ecosystem services in urban areas. Curr Opin Environ Sci 14:101–108

Team GreenKeys (2008) GreenKeys @ your city—a guide for urban green quality. IOER Leibniz Institute of Ecological and Regional Development, Dresden

Grimm NB, Faeth SH, Golubiewski NE, Redman CL, Wu J, Bai X, Briggs JM (2008) Global change and the ecology of cities. Science 319:756–760

Grunewald K, Bastian O (eds) (2015) Ecosystem Services—Concept, Methods and Case Studies. Springer, Berlin, Heidelberg, New York

Haase D, Frantzeskaki N, Elmqvist T (2014a) Ecosystem services in urban landscapes: practical applications and governance implications. Ambio 43(4):407–412

Haase D, Larondelle N, Andersson E, Artmann M, Borgström S, Breuste J, Gomez-Baggethun E, Gren A et al (2014b) A quantitative review of urban ecosystem service assessments: concepts, models, and implementation. Ambio 43(4):413–433. doi:10.1007/s13280-014-0504-0

Hansen R; Heidebach M, Kuchler F, Pauleit S (2012) Brachflächen im Spannungsfeld zwischen Naturschutz und (baulicher) Wiedernutzung. In: Bundesamt für Naturschutz (ed) BfN-Skripten 324. Bonn

Huang D, Huang Y, Zhao X, Liu Z (2017) How do differences in land ownership types in China affect land development? A case from Beijing. Sustainability 9(1):123. doi:10.3390/su9010123

Kraas F (2008) Megacities—our global urban future. In: Derbyshire E (ed) The International Year of Planet Earth. London, pp 108–109

Kowarik I (2005) Wild urban woodlands: Towards a conceptual framework. In: Kowarik I, Körner S (eds) Wild urban woodlands. New perspectives for urban forestry. Springer, Heidelberg, pp 1–32

Kuper A, Kuper J (eds) (1996) The social science encyclopedia, 2nd edn. Routledge, London

Lennon M (2015) Green infrastructure and planning policy: a critical assessment. Local Environ 20(8):957–980

Lindfield M, Steinberg F (eds) (2012) Green cities. Mandaluyong City. Asian Development Bank, Philippines

MEA—Millennium Ecosystem Assessment (2005) Ecosystem and human well-being: Scenarios, vol 2. Island Press, Washington

Milchert J (2003) Visionen für die Landschaftsarchitektur. Garten + Landschaft 11:23–25

MoHURD—Ministry of Housing and Urban-Rural Development of the People's Republic of China (2002) The standard classification of urban green space in the People's Republic of China (in Chinese). http://www.tzzfj.gov.cn/art/2016/6/22/art_8067_337107.html. Accessed 12 Dec 2016

Pauleit S, Liu L, Ahern J, Kazmierczak A (2011) Multifunctional green infrastructure planning to promote ecological services in the city. In: Niemelä J (eds) Urban ecology: patterns, processes, and applications. Oxford University Press, New York, pp 272–285

Richter G (ed) (1981) Handbuch Stadtgrün: Landschaftsarchitektur im städtischen Freiraum. BLV Verlagsgesellschaft, München
Ridder B (2008) Questioning the ecosystem services argument for biodiversity argumentation. Biodivers Conserv 17:781–790
Rittel K, Bredow L, Wanka ER, Hokema D, Schuppe G, Wilke T, Nowak D, Heiland S (2014) Green, natural, healthy. The potential of multifunctional urban spaces. Federal Agency for Nature Conservation (BfN) (publisher), Bonn
Rouse DA, Bunster-Ossa IF (2013) Green infrastructure: a landscape approach. APA Planning Advisory Service
Schröder A, Arndt T, Mayer F (2016) Naturschutz in der Stadt – Grundlagen, Ziele und Perspektiven (Nature conservation in the city—basic principles, aims and perspectives). Natur und Landschaft 91(7):306–313
SDG—Sustainable Development Goals (2015) Sustainable development goals: FACT SHEET. http://www.un.org/sustainabledevelopment/wp-content/uploads/2015/08/Factsheet_Summit.pdf. Accessed 10 Feb 2016
Seto KC, Reenberg A (2013) Rethinking global land use in an urban era. The MIT Press, Cambridge
Siemens AG (2012) The green city index. www.siemens.com/greencityindex. Accessed 12 Dec 2016
TEEB—The Economics of Ecosystems and Biodiversity (2010) The economics of ecosystems and biodiversity: ecological and economic foundations. In: Kumar P (ed) Earthscan, London, Washington
TEEB DE—Naturkapital Deutschland (2016) Ökosystemleistungen in der Stadt – Gesundheit schützen und Lebensqualität erhöhen. In: Kowarik I, Bartz R, Brenck M (eds) Technische Universität Berlin, Helmholtz-Zentrum für Umweltforschung – UFZ, Berlin, Leipzig
UNDP—United Nations Development Programme (2012) The future we want: biodiversity and ecosystems—driving sustainable development. United Nations Development Programme Biodiversity and Ecosystems Global Framework 2012–2020, New York
UNICEF—United Nations Children's Fund (2012) The State of the World's Children
UN—United Nations (1992) Convention on Biological Diversity (CBD)
UN—United Nations (2014a) Convention on Biological Diversity. Concluded at Rio de Janeiro on 5 June 1992
UN—United Nations (2014b) World Urbanization Prospects: The 2014 Revision, Highlights (ST/ESA/SER.A/352), New York
URGE Team (2004) Making Greener Cities—a practical guide. UFZ-Bericht Nr. 8. Stadtökologische Forschungen 37. UFZ Leipzig-Halle, Leipzig
WBGU—Wissenschaftlicher Beirat der Bundesregierung Globale Umweltveränderungen (2016) Der Umzug der Menschheit: Die transformative Kraft der Städte. Wissenschaftlicher Beirat der Bundesregierung Globale Umwelt veränderungen, Berlin
Wenzel J, Schöbel S (2001) Weite Felder: Jenseits einer Theorie zukünftiger Landschaftsarchitektur. DAB – Deutsches Architekten Blatt 7
Wittig R (1999) Was soll, kann und darf der Naturschutz in der Stadt? Geobot. Kolloq. 14:3–6, Frankfurt
Wu J (2014) Urban ecology and sustainability: the state-of-the-science and future directions. Landscape Urban Plann 125:209–221
Wurster D, Artmann M (2014) Non-monetary assessment of urban ecosystem services at the site level—development of a methodology for a standardized selection, mapping and assessment of representative sites. Ambio 43(4):454–465. doi:10.1007/s13280-014-0502-2

Chapter 2
Conceptual Framework

Jürgen Breuste, Junxiang Li and Karsten Grunewald

A conceptual approach for greener cities must always be based on recognized actual and expected challenges. A first reaction is to develop strategies to deal with these challenges. These strategies have to be conceptualized in clear, targeted concepts. This is the main idea of this chapter and of the book as a whole. To do this for two very different countries, different in size, population, and urbanization level (Sect. 2.3), is a challenge of its own. The need to come to positive results may be much more challenging in China than in Germany. But the strategies can be readily compared, and the two sides can learn from each other about the efficiency of strategies and concepts.

J. Breuste (✉)
University Salzburg (AUT), Salzburg, Austria
e-mail: juergen.breuste@sbg.ac.at

J. Breuste
East China Normal University Shanghai, Shanghai, China

J. Li
East China Normal University and Shanghai Key Laboratory of Urbanization and Eco-Restoration, Shanghai, China
e-mail: jxli@des.ecnu.edu.cn

K. Grunewald
Leibniz Institute of Ecological Urban and Regional Development, Dresden, Germany
e-mail: k.grunewald@ioer.de

K. Grunewald et al. (eds.), *Towards Green Cities*, Cities and Nature,
DOI 10.1007/978-3-319-58223-8_2

2.1 Challenges Developing Greener Cities in China and Germany

Jürgen Breuste and Junxiang Li

Two countries, China and Germany, different in population, size, and culture, share comparable challenges from urbanization. Also, these challenges are different in their dimensions but not very different in their main aspects (Table 2.1). The focus here should not be on challenges which are either very specific to only one country, like urban shrinkage in several German cities or handling of de-industrialization, or very regional within one of the countries, but rather on those which are common to both countries. This includes also the public, planning and administrative reaction to those challenges and the methods used to overcome them.

Concentrating more than 50% of the world's population on less than 3% of the Earth's surface, urban areas provide tremendous opportunities for sustainable land use (Box 2.1). The development of urban areas, including growing infrastructure, economic development, and quality of life, is to be brought into harmony with the protection of nature and the environment. Many municipalities and initiatives worldwide are pursuing 'green' urban development (Fig. 2.1). The concepts that serve as a basis for this are diverse. Transparent, indicator-based evaluation systems are necessary to ensure that planning and action do indeed lead to increased sustainability and to a higher quality of life of the population in cities (Box 2.2).

Box 2.1 More individual living space forces urban growth and resource consumption in Germany and China

In ***Germany*** the trend to bigger apartment sizes per capita is increasing. Between 1998 and 2013 the consumption of apartment space increased from 39 to 46.3 m^2, with pollution constant or even decreasing. The reasons are economic welfare imposing more individual household structures (one- or two-person apartments) and the increasing demand for individual space in apartments (BiB 2013).

Some 50 million of ***China***'s 230 million urban households still live in substandard quarters in crowded conditions. It is estimated that China will need to build 10 million apartments a year until 2030 (Orlik and Fung 2012). Currently, China's average per capita living space is already 32.9 m^2; in Shanghai it is 24.2 m^2 (GBtimes 2015). This trend in individual space resource consumption results in higher CO_2 emissions, more traffic, more pollution, more energy and material use, and many other important factors relevant for sustainable urban development.

Table 2.1 Different conditions of urbanization and ecological urban development (simplified overview)

	Germany	China
Urbanization rate	75% (2014)	54% (2014)
Economic growth rate	Low economic growth, tendency decreasing	High economic growth, tendency decreasing
Urban growth	Slow urban growth through changes of internal urban structures, but still growing urban sprawl and urban land consumption of rural land	Extreme urban growth with big regional differences and the target to reach the urbanization degree of developed countries, and a ban (statutory, but patchily enforced) on converting agricultural land into built-up area
Management of urbanization	Locally, regionally but also nationally steered urban development (e.g. national/regional/local strategies for different challenges like sustainability, climate change, loss of biodiversity, brownfield development), decisions based on competition for economic resources and governmental moderation. Competition between cities and surrounding communities	Centrally steered urban development policy, competition for centrally distributed financial support, local and regional economic competition. Building of new cities and changing of old cities by radical transformation and building of new urban structures (infrastructure, housing stock, open spaces, industry, etc.), but also conservation of cultural treasures. Recently, this phenomenon has slowed down
Ecological principles of urban development	At the urban district level with involvement of citizens and NGOs, good examples have been realized based on urban ecological knowledge for decision-making and urban design	Centrally managed strategy of building new cities as eco-cities without involvement of citizens, based only on fragmented urban ecological knowledge for urban design. But this is now changing with the central and/or local government's efforts to input sound ecological knowledge into decision making as well as urban planning and design
Green policy	Different approaches for green space development policies in cities: national (e.g. 'Whitebook Urban Green', national strategies), regional (e.g. regional plans), local (e.g. landscape, master plans for green, green/biodiversity strategies), supporting programs, etc.	Powerful green space development policies and urban green extension in many cities based on high financial input in land, construction and management. Urban ecological space construction and environmental protection are more economically and administratively oriented, as opposed to ecological orientation

(continued)

Table 2.1 (continued)

	Germany	China
Development of urban nature	Protection of existing urban nature, especially urban forests and wetlands for recreation and biodiversity. Partly development of new urban nature on brownfields (urban wilderness), in urban reconstruction processes. Development of green networks to connect green spaces and biotopes in the city	Development of new urban nature, especially planting urban forests and new design of urban wetlands for recreation and biodiversity

Fig. 2.1 Green can be integrated into urban patterns: **a** Urban green in a residential area in Beijing, **b** Urban green in a residential area in Dresden, **c** Roads without green cover and air pollution in Beijing, **d** Green street/parking place in Beijing. © Jürgen Breuste

Box 2.2 Ten challenges to develop greener cities in China and Germany (cf. Breuste et al. 2016)

1 … Growing urban land consumption

The use of land for building activities and the resource utilization in and around agglomeration centers are increasing. This is connected with increasing resource consumption of mostly fertile agricultural land and other nature resources.

2 … Low-density urban expansion
The improving personal living conditions and the individualization in cities entail a preference for low-density housing and more individual space. The demand is high in Germany and growing in China.
3 … Unhealthy urban living conditions due to air and water pollution and noise
Most unhealthy urban living conditions are caused by air and water pollution, too little green space and noise. Especially urban dwellers in China suffer under these conditions.
4 … Climate change
Climate change will aggravate the thermal conditions in urban areas and will be a health risk for sensitive and exposed urban dwellers.
5 … Uneven distribution of urban green
In most of the cities the urban green is not equally distributed and accessible for all to benefit from it.
6 … Exclusion of people from decision-making about their neighborhoods
Only accepted local decision-making is sustainable. Most people want to participate in decision-making about their surroundings. They know best which urban green is missing and where. In many ways ordinary people are, especially in China, still not integrated in the decision-making.
7 … Unused existing scientific and practical knowledge
Many new development projects are failing to integrate the highest standards of available ecological knowledge on buildings and green space organization. Especially in China with its huge urban development everywhere, the ecological standards for building and for open spaces are mostly still not implemented.
8 … Reduced public budget for local tasks
In many communities, reduced budgets do not allow for the development of green spaces with high follow-up maintenance costs. This is the case especially in Germany, but China is also faced with this.
9 … Urban dwellers lost contact with nature
Contact with nature is necessary for health, for a general understanding of nature, and for benefiting from nature. Most urban dwellers have already lost close contact with nature but would like to develop it again.
10 … Existing cities are not sustainable
The current urban form, internal spatial quality, energetic performance, traffic organization, etc. are not sustainable and resilient against future challenges.

Most of the challenges cannot be met by technical solutions alone. These must be accompanied by a reduction of resource-consuming behavior, a long-term process of acceptance and education. The potentials of ecosystems, often called 'nature-based solutions', can be used in support. The problems for which urban nature can contribute to a 'solution' have to be identified carefully, without overestimating urban nature's capacities, but also not underestimating and ignoring them. In many cases all three aspects, technical, behavioral, and nature-based, will have to be included in order to best meet the contemporary and future challenges to building resilient cities and towns.

There are several scientific challenges for sustainable urban development integrating urban nature:

- Urban nature and its role in providing ecosystem services is intensively being discussed internationally (Breuste et al. 2013; Kabisch et al. 2015; Von Döhren and Haase 2015; Hansen and Pauleit 2014; Hansen et al. 2015). The analysis and assessment of urban ecosystem services and of urban biodiversity needs to be scientifically sound and practice-oriented. There is an urgent need to integrate the concept of urban nature into urban planning and management (Pauleit et al. 2011; Hansen and Pauleit 2014; Hansen et al. 2015).
- The contribution of the different green and blue spaces which are part of the urban green infrastructure needs to be evaluated with respect to their contribution to people, biodiversity, and climate change adaptation (Gill et al. 2007; Fryd et al. 2011; Breuste and Artmann 2014; Loibl et al. 2014).
- There can be trade-offs between urban ecosystem services and biodiversity. There is a need to investigate how they are mutually dependent within different urban contexts (Breuste et al. 2013; Wang et al. 2016).
- The scales of investigation and implementation are a special problem of analysis and assessment (Andersson et al. 2015). The local-level provision of ecosystem services is important but needs to be embedded into strategic frameworks for developing green infrastructure at urban and regional scales (Pauleit et al. 2011).
- Development of better relationships between urban and rural areas is crucially important in both the European and the Chinese context (Spiekermann et al. 2013). Scale mismatches between service-providing units and units of decision-making require particular attention (Borgström et al. 2006).
- Urban waters are important parts of the urban green-blue infrastructure. Their protection and reestablishment in order to let people profit from urban nature is currently ongoing in many cities (Sect. 3.3.2).
- Soil sealing reduction is a general target in many cities. Several methods to improve this are being discussed (Pauleit and Breuste 2011; Artmann 2014).

2.2 Strategies and Concepts

Juliane Mathey, Stefanie Rößler, Anne Seiwert, Jiang Chang, Tinghao Hu, Suili Xiao and Qiaoqiao Xu

In this study, the 'green city' as one concept for urban development refers to a city which fosters urban biodiversity to form green infrastructure, which might provide ecosystem services based on nature-based solutions to tackle urban challenges.

Strategic and conceptual approaches can help to define the aims for urban development and the future spatial structure of a city (Yu et al. 2011). If such approaches for the development of entire urban open space systems are in place, they can serve as a firm base for decision-making (Hansen and Pauleit 2014) and can thereby foster the development of 'green cities'.

To tackle the above-mentioned challenges (Sect. 2.1) and to counteract the problems connected with them, a series of policies, concepts, and strategies referring to different planning levels exist: national, regional, and local (entire city level and neighborhood level).

Concepts on the one hand provide ideas for the urban structure and on the other hand contribute to achieving a concrete aim, the implementation of 'green cities'. For the implementation of policies, models, and concepts, generally strategic planning or strategic approaches are required. Bryson (2004) defines strategic planning in general as "a disciplined effort to produce fundamental decisions and actions that shape and guide what an organization (or other entity) is, what it does, and why it does it." It consists of a set of concepts, methods, and instruments. Bryson (2004) further emphasizes that strategic planning is a tool that supports thinking, acting, and learning in a holistic and organizational context. Strategies can be seen as informal planning instruments. Strategic approaches can aim at the spatial structure of a city, at the quality and function of urban green space systems and at the implementation process.

In this chapter, the following main approaches to foster the development of a 'green city' will be introduced: policies and political strategies (Sect. 2.2.1), concepts for cities (Sect. 2.2.2) and municipal green strategies (Sect. 2.2.3).

2.2.1 Political Strategies

Political strategies try to advance political ideas (Schröder 2000). Currently the concept of a 'green city' is being promoted quite prominently through a number of political strategies on different spatial and political levels.

Global strategies

The 17 Sustainable Development Goals (SDGs) adopted in 2015 are set to take into account all aspects of sustainable development—the ecological, economic, and social dimension—more than has hitherto been the case. For example, one of the core elements of Goal 11 refers to universal access to safe, inclusive and accessible, green and public spaces in particular for women and children, older persons, and persons with disabilities (UN 2015). Since 1978, UN-Habitat has been specifically setting the focus on urban spaces, aiming at promoting the development of socially and ecologically sustainable settlements (UN-Habitat 2012). At the Habitat III Conference in October 2016, the New Urban Agenda was passed, which is meant to serve as a political guideline for urban development over the next two decades. Ecological sustainability is one issue among manifold other goals for urban settlements. It should be ensured *inter alia* by protecting, conserving, restoring, and promoting ecosystems and biodiversity to allow healthy lifestyles in harmony with nature, urban resilience, and mitigation of and adaptation to climate change. Besides, also safe, inclusive, accessible, and green public spaces in the form of multifunctional areas should be promoted (UN 2016).

Also at the international level, the Convention on Biological Diversity (CBD) is seen as a key instrument for fostering and protecting biological diversity (Harrop and Pritchard 2011). So far, 196 Contracting Parties (as of September 2016, CBD 2016), including China and Germany, have ratified this convention.

Strategies for Europe and Germany

Apart from these global strategies, the diverse ecological characteristics of a country and its specific social, political, and economic factors involved in the protection and sustainable exploitation of biological resources require national and subnational responses (Harrop and Pritchard 2011). The CBD serves as a conceptual framework for respective national strategies. In Germany, the 'National Strategy on Biological Diversity' adopted in 2007 (BMU 2007) meets these requirements by demonstrating the German contribution to the conservation of biodiversity at the national and international level, providing guidance for different actors (Stiehr 2009). One of its concrete visions relates to urban landscapes and aims *inter alia* at increasing green spaces in settlements, e.g., through unsealing or ecological upgrading of residential districts.

The German 'Greenbook on Urban Green' summarizes the state of knowledge on urban green development (BMUB 2015). Based on this, a 'Whitebook' has been released in May 2017, which presents concrete recommendations for action and implementation in terms of ten actions fields (BMUB 2017). Since 2011, the 'EU Biodiversity Strategy to 2020' aims at reversing biodiversity loss and accelerating the EU's transition towards a resource-efficient, environmentally friendly economy (EU COM 2011). Against the background of a highly fragmented landscape, the 'EU Green Infrastructure Strategy' published in 2013 serves to promote an EU-wide green infrastructure. Establishing a green infrastructure should foster the preservation and improvement of ecosystems and the services they provide (EU COM 2011). Furthermore, green infrastructure is perceived as an opportunity to integrate biodiversity issues into other policy sectors (EU COM 2013).

The development of urban green spaces is strongly related to issues of urban land use. The National Sustainability Development Strategy of Germany claims to reduce the land take for settlements and traffic infrastructure to less than 30 ha per day by the year 2030 (Die Bundesregierung 2016). This aim is to be reached *inter alia* by avoiding urban sprawl and fostering compact cities (see Table 2.3). The latter goal, which is mainly addressed by infill development in existing urban structures, has an influence on the provision of green spaces in cities. To deal with this conflict, the approach of so-called 'dual inner development' has been introduced to ensure high-quality green space concurrent with building activities in urban areas (Kühnau et al. 2016).

Strategies for China

In the case of China, after it ratified the Convention on Biological Diversity in 1992, the national government released the China Biodiversity Conservation Action Plan in 1994 and updated it to the China National Biodiversity Conservation Strategy and Action Plan (2011–2030) in 2010 to guide conservation and protection work at the national level.[1] To address the importance of biodiversity and green space in urban areas, the State Standard for Garden Cities[2] released by the Ministry of Environmental Protection (MEP) in 2000 and the Notice on Enhancing the Work on Protecting Urban Biodiversity[3] by the Ministry of Housing and Urban-Rural Development (MoHURD) in 2002 set clear indicators and targets to evaluate biodiversity and the condition of green spaces, in order to better preserve and restore them for ecological and cultural services and socioeconomic benefits.

Most recently, China's 13th Five-Year Plan (2016–2020) on National Ecological Protection,[4] a national legislative document to set the direction, principles, baselines, and targets and map out the strategies for ecological conservation and protection in China for the next 5 years, mentioned protecting urban biodiversity and restoring urban green space as key complementary factors to expand ecological services. In addition, the document National Planning of Urban Ecological Protection and Construction (2015–2020)[5] jointly published by MoHURD and MEP in December 2016 proposed to improve the living environment, strengthen urban biodiversity protection, and restore urban eco-environments. In order to implement the targets set in the national plan, provinces and cities usually make their own Five-Year Plans, listing specific targets including the urban park area per capita, etc.

China's Plan for National Economic and Social Development, also called the Five-Year Plan, is the leading plan of national economic and social development in a certain period. It plays a leading role, in contrast to a specific implementation plan. Since 1953, China has adopted 13 Five-Year plans, which covered economic and social development, industry, IT, ecological and environmental protection, etc. Regarding green space development the plans mainly act as steering wheel and set the aggregate indicators on national scale.

'Ecological civilization' is one of the new government strategies in the 13th Five-Year Plan. The core and essence of ecological civilization is to maintain the natural ecological balance and realize harmony between human beings and nature. The specific framework includes: strengthening ecological development, regulating land use, developing main functional zones, strengthening the construction of ecological society, promoting citizens' awareness of ecological development, establishing an ecological civilization system, and setting up a system of national parks.

---

[1]https://www.cbd.int/doc/world/cn/cn-nbsap-v2-en.pdf.

[2]http://www.mohurd.gov.cn/lswj/tz/201012502.doc.

[3]http://www.mohurd.gov.cn/zcfg/jsbwj_0/jsbwjcsjs/200611/t20061101_157066.html.

[4]http://www.gov.cn/zhengce/content/2016-12/05/content_5143290.htm.

[5]http://www.mohurd.gov.cn/wjfb/201612/t20161222_230049.html.

Environmental protection and ecological control program at the national level is a generic term for a series of plans that play an active role in promoting green space development. This category of plans further extends and refines the relevant provisions of green space development and environmental protection in the Plan for National Economy and Social Development, so as to provide ideas and guidance for green space development. These plans and outlines are mainly issued by the State Council, the Ministry of Environmental Protection, the Ministry of Housing and Urban-Rural Development, the State Forestry Administration, etc. Some environmental protection and ecological control programs issued in recent years that take an active part in promoting green space development are listed in Table 2.2. Nevertheless, China has not yet issued a special plan with the theme of green space or green infrastructure systems at the national level.

2.2.2 Urban Concepts

The communication and finally agreement of different stakeholders on one vision of urban green space development might support the coordination of different interests and the implementation of concrete measures of urban development (Becker 2010; WBGU 2016).

So-called 'Leitbilder' concerning visions, guiding principles, overall concepts or objectives for urban development are used in different contexts (Rößler 2010) to transport ideas for the 'city of the future' in reaction to changes of society, environment, and economy (Kuder 2008). Related to their different origins, the rationales, scope, relevance, and contents differ (Fürst et al. 1999). Three main types can be distinguished (Sieverts 1998):

- 'Urban Utopias' are characterized by symbolic dimensions defining ideal urban frameworks, integrating spatial, societal, and socioeconomic aspects (e.g., Garden City, Howard 1898);
- 'Structural models' focus on urban form, physical structures, land use, and the spatial form (e.g., polycentric urbanization models, compact city, green rings, Sect. 4.2);
- 'Slogans' with a strategic orientation, mottos and catchwords, reproduce more complex and manifold ideas and visions. They are often used for marketing purposes or in competitions (e.g., Vancouver 'Greenest City'; 'Golden/Garden city Wenjiang', Sect. 4.3.7).

Most urban strategies combine selected aspects of more than one of these types to form general but also individual urban strategies.

Green spaces have been addressed explicitly in some of these approaches. Since the beginning of the twentieth century, particularly social, health, and economic functions of urban green spaces have been considered (e.g., Garden City, Howard

1898; organic decentralism, Zhao 2011). Currently, green spaces and their different functions (ecosystem services) are a central and natural element in urban concepts (Breuste et al. 2013; Sondermann and Rößler 2016).

Table 2.2 Plans and programs for promoting green space development at the national level in China

Plans	Departments	Planning Period	Requirements for Green Space Development
Plan for Promoting 'Ecological Civilization' development	State Forestry Administration	2013–2020	It proposes specific requirements of urban forestry and aesthetic rural development
National Main Function Zones Planning	State Council; Central People's Government	Issued in 2010 and planned to be realized in 2020	National territory is designated in zones of so-called 'optimization development', 'key development', 'limited development', and 'prohibited development'. The plan emphasizes that green space protection is a vital requirement of sustainable territorial development
National Ecological Protection and Construction Planning	National Development and Reform Commission, and other 12 departments	2013–2020	It proposes specific suggestions for improving urban ecology, including: urban green system planning, urban forest and country park construction, urban heat island control, urban water quality management, urban vertical greening and low elevation greenbelt construction
National Forestation Program	State Forestry Administration	2011–2020	It puts forward requirements and measures on urban and rural afforestation, green channel and green network construction, restoration of post-mining areas
Ecological Function Area Planning	Ministry of Environmental Protection; Chinese Academy of Sciences	From 2015	According to the natural conditions and ecosystem services potential, so-called 'ecological function areas' at a national scale can be categorized into 3 major classes (ecological regulation, product provision, and human security); there are 242 such areas in China

The variety of different concepts and the speed with which new, additional or even contradictory ones are being developed are continuously increasing. The different approaches "… often appear to be used interchangeably by policy makers, planners and developers" (De Jong et al. 2015). Nevertheless, each of them has conceptual cores and perspectives which make them distinguishable, though also partly overlapping.

The most widespread (and by now mainstream) concept is the 'sustainable city'. Originally, sustainable development was defined as development that meets the needs of the present without compromising the ability of future generations to meet their own needs (Brundtland Commission 1987). In 1991 the United Nations Centre for Human Settlements (UNCHS) defined a sustainable city as one "where achievements in social, economic and physical development are made to last" (UN-Habitat 2002). As this definition has been considered too general, particularly lacking ecological aspects (Rees 1992), there has been an intense debate about its aims since the early 1990s (DESA 2013). Subsequently, the Rio Declaration on Environment and Development (UN 1992) integrated the economic, social, environmental, and governability dimensions of sustainability. For the European context it has been stated in the so-called 'Leipzig Charter' that "all dimensions of sustainable development should be taken into account at the same time and with the same weight. These include economic prosperity, social balance and a healthy environment. At the same time attention should be paid to cultural and health aspects" (EU 2007). Concepts and practices of 'sustainable urbanism' have gained growing international popularity since the early 2000s and entered mainstream policy as a consequence of the forceful combination of global climate change concerns and a rapidly urbanizing world population (Huang et al. 2015). Meanwhile this term is accompanied, specified or even replaced by numerous 'sister terms' as for example 'eco-city', 'green city', 'low-carbon city', 'resilient city' or 'compact city' (Joss 2015; de Jong et al. 2015; Table 2.3).

The idea of the 'eco-city', i.e., an 'ecological city', developed in the mid-1970s, with the aim of (re)constructing cities in balance with nature (de Jong et al. 2015). It was defined as a city "one built according to the principles of living within the means of the environment" (Register 1973). The approach of focusing on ecological principles of urban development also followed the ideas of UNESCO's Man and Biosphere (MAB) Program (Zhao 2011). Over time, a broad understanding and a variety of meanings and interpretations has been developed "of which the ecological may be the main, but certainly not the only, one" (de Jong et al. 2015). The term 'eco-city' is increasingly used to describe sustainable urban development, focusing on ecological aspects such as resource efficiency, low-carbon development or waste reduction. Concrete projects to build eco-cities are initiatives that variably promote and pursue sustainable development in relation to urban infrastructure, services, and community at district, town, or metropolitan levels (Joss 2015). In China, more than 500 'ecological cities (counties)' and demonstration projects are under construction. Flagship eco-city projects such as the Sino-Singapore Tianjin

Table 2.3 Overview of concepts related to 'eco-city' (Joss et al. 2011)

Sustainable city	Synonymous with 'eco-city/town'. The UN-Habitat Sustainable Cities Programme has been promoting this concept since the early 1990s
Sustainable community	Synonymous with 'eco-community'
Smart city	Used to emphasize hi-tech aspects of development—smart energy grids, IT networks, and related efficiencies in utility and service provision
Slim city	World Economic Forum knowledge transfer initiative to encourage cities to increase efficiency across a variety of sectors, e.g. energy, transport, construction work
Compact city	Use of this term typically implies an opposition to urban sprawl. It is an influential urban design concept whose guiding principles include high residential density and the discouragement of private car use
Zero energy city/zero net energy city	Uses no more energy than it is able to generate locally. This is achieved through a combination of measures to reduce power consumption and the introduction of new renewable energy sources
Low-carbon city	The reference to carbon (in this and the following two terms) may reflect national aspirations to create 'low-carbon economies'—often as part of policies designed to mitigate climate change. The focus is on the physical aspects of cities: energy, transportation, infrastructure and buildings. 'Carbon' is sometimes used as shorthand for all greenhouse gases
Carbon-neutral city/net zero city	Similar to 'low-carbon city'—except defined more strictly as a city which offsets carbon/greenhouse gas emissions such that its net emissions are zero
Zero-carbon city	More specifically still, a city which produces no greenhouse gases and is run exclusively on energy from renewable sources
Solar city	May have a relatively narrow focus on replacing fossil fuels with solar energy, and is in some cases limited in its ambitions. The Indian government's Solar Cities program aims to reduce conventional energy use by 10%, with solar energy being part of a mix of renewable energy sources to be promoted
Oekostadt/Ökostadt	As well as being a direct German translation of the term 'eco-city', 'Ökostadt' also refers more specifically to a series of Austrian, German and Swiss towns and cities which declared their intention to introduce principles of sound environmental management and sustainable development in the 1990s—often as part of an Agenda 21 program
Transition town	The 'Transition Town' movement, which originated in the UK and Ireland, is a growing international phenomenon. Transition Town activities are typically organized at grass-roots level rather than embedded in policies. The aim is to build up local communities' social and environmental resilience to the effects of climate change and fossil fuel shortages—both of which are assumed to be inevitable in the future
Eco-municipality	The label 'eco-municipality' describes a local authority which has adopted a particular series of values related to environmental and social sustainability, to guide policy making. The movement is most strongly associated with Sweden (where it has its roots in the 1980s), but has also recently gained ground in the USA

project were established (Shepard 2015). Meanwhile, 'eco-city' works as an "umbrella term encapsulating a variety of concepts, models and practices which aim to further urban sustainability at neighborhood, city or city-regional levels" (Joss et al. 2011, Table 2.3).

Currently the concepts of 'eco-cities' and 'green cities' (Beatley 2012) as starting points for sustainable development are emerging and spreading. Nevertheless, the UN points out that it is "… important to understand cities' sustainability as a broader concept which integrates social development, economic development, environmental management and urban governance, which refers to the management and investment decisions taken by municipal authorities in coordination with national authorities and institutions" (DESA 2013). Following the Sustainable Development Goals of the UN, the 'New Urban Agenda', adopted at the United Nations Conference on Housing and Sustainable Urban Development (Habitat III) in 2016 (Sect. 2.2.1) formulates a broad concept of future cities and human settlements (Habitat III 2016). One issue, besides others, is that cities should "protect, restore, and promote their ecosystems, water, natural habitats and biodiversity, minimize their environmental impact, and change to sustainable consumption and production patterns".

In this book, we use the term 'green city' to refer to a city which focuses on green spaces, nature, and biodiversity as crucial elements of a sustainable city (Sect. 1.2).

2.2.3 Municipal Strategies for Urban Green Spaces and Biodiversity

Even though urban green spaces and urban nature, due to manifold challenges (see above), are of higher relevance than ever, land use decisions in cities are often made against them (Kowarik et al. 2016). Against this background, strategic approaches on the municipal level are useful tools to strengthen the position of urban green spaces and biodiversity and can help to improve urban green space systems by defining overall aims, providing spatial frames, and serving as firm bases for decision-making (Yu et al. 2011; Hansen and Pauleit 2014).

Municipal Green Space Strategies

"An Urban Green Space Strategy is a strategic document with a long-term perspective that should fall within the compass of a city's development policies and integrate well with other policies. It deals with all urban green spaces, regardless of type or ownership. To be most effective and for the greatest efficiency the strategy should be integrated into the planning system of the city. Such a strategy confronts the present situation of green spaces (with all problems, conflicts, potentials and needs) and the future collective vision and goals. It covers all aspects and subjects dealing with green spaces management and development. As a result it provides basic and enduring development proposals, tasks and actions for implementation that are needed to assure the realization of set visions and goals" (GreenKeys Team 2008). By addressing all types of green spaces or biotopes, urban green space strategies have the potential to provide a basis for the protection and promotion of biodiversity (habitats and species). Urban green space strategies can address all aspects of the internal and external interactions that are important in the context of green spaces. They can address

- the quantity (minimum green standard values and benchmarks, e.g., m^2 green spaces per capita, Sect. 3.7),
- the spatial structure of a city and the green space system (density, land use patterns: e.g., green rings like in Leipzig, Cologne and also in Chinese cities or the vision "Dresden as a compact city in an ecological network" (Sect. 4.4.6); green fingers; connecting built-up with green and open spaces; accessibility of green: e.g., distance from residential area),
- the quality and function of urban green space systems (e.g., priority areas, green space types, aims for biodiversity and ecosystem services, green spaces for play and sports , Sect. 4.2),
- the implementation process (e.g., planning, building, management of urban green spaces, public participation, formal and informal instruments, programs, e.g., for greening of walls, roofs, backyards , Sect. 5.1 and 5.2) and
- the monitoring, e.g., an evaluation of implementation processes (success and failures).

In China, following the guidance of national laws, local governments, at both the provincial and city level, usually set up the regulations on urban greenery, wildlife protection, and historical areas, etc., to carry out urban green and biodiversity works; meanwhile, many cities have made or are making specific urban biodiversity protection plans under their own urban green system planning and are actively participating in the national initiatives to build a 'Garden City', which regards urban biodiversity as a crucial factor for its development. Since 1992, almost half of all Chinese cities (310 cities) and 212 counties have been nominated by MoHURD as 'Garden Cities' or 'Garden Counties'.[6]

However, cities take different approaches to improve their green space and strengthen local biodiversity; some of them are more focused on protecting the existing species and green spaces, some highlight the importance of enriching the species and restoring the environment; some are only about the plants, some also include animals (Hu 2011). In order to further evaluate the efforts and accelerate the progress on urban ecological protection and restoration, the document Standards on Classification of National Eco-Garden Cities was published by MoHURD with a

[6]http://www.mohurd.gov.cn/zxydt/201602/t20160201_226501.html.

comprehensive evaluation index system and multidimensional performance assessment methodology. The index system covers several aspects with clear and ambitious targets such as local species ratio in urban green areas, resident satisfaction percentage, storm water recollection ratio, etc., and the performance assessment methodology includes local engagement, expert field studies, and third-party verification. On 29 January 2016, MoHURD designated seven cities as 'National Eco-Garden Cities', Xuzhou, Suzhou, Kunshan, Shouguang, Zhuhai, Nanning and Baoji, to recognize their achievements: For example, to restore the urban ecosystem and improve urban ecosystem services, Baoji City has carried out exemplary projects to renovate urban mountain environments, such as Beipo and Nanyuan; Nanning City has successfully recovered the urban water systems, such as Keli River and Liangqing River, etc.; Shouguang City has systematically restored the urban mining areas; to build up the urban greenery system and coordinate regional development, Kunshan City has formed a new urban system with integration of water, road and green; Zhuhai also has built 298 community parks to ensure that the citizens can walk to a park within 10 minutes; and Shuzhou city has established a well-connected ecological network.[7]

Municipal Biodiversity Strategies

For effective and efficient protection and development of local biodiversity, it is essential to proceed strategically. For this reason, a number of German municipalities are planning or developing or have even already passed biodiversity strategies. Through these strategies it is possible to systematically record, describe, and negotiate objectives and activities to promote urban nature with respect to legal, economic, planning, and ecological aspects. Municipal biodiversity strategies do not replace instruments like environmental reports, landscape plans, species protection programs or action plans, but rather place them in a common context. The resulting benefit is not only a document, but also the process of elaboration and implementation, providing the chance to develop common ideas for local nature and biodiversity together with citizens, nature conservation organizations and other stakeholders. This contributes to a better understanding and appreciation of biodiversity by residents. By having such a strategy approved by the city council, a high level of political commitment and support can be ensured (Herbst 2014). At present (year 2016), 12 German municipalities have a biodiversity strategy.

In China, all provinces (23) have biodiversity strategies. There are three municipalities which have biodiversity strategies, namely Shanghai, Tianjin, and Chongqing (Chinese versions available under http://www.doc88.com/p-5843925970015.html (Shanghai), http://www.doc88.com/p-7846250170700.html (Tianjin), http://www.doc88.com/p-5317308281162.html (Chongqing)). Also some prefecture-level cities like Kunming and Haikou and some county-level cities like Dujiangyan have biodiversity strategies.

---

[7]http://www.mohurd.gov.cn/zxydt/201602/t20160201_226501.html.

2.3 Aspects of City Characteristics in China and Germany

Karsten Grunewald, Wei Hou, Gaodi Xie and Jürgen Breuste

Cities differ in many ways, from age and architecture to climate, ecology, economy, and culture (Liu et al. 2016). It should be noted that Chinese and German cities can be compared physically (Tables 2.4 and 2.5), but regarding the sociocultural conditions and values this is very difficult. The frame for the following explorative case study was selected with consideration of data availability and own experiences of the author team.

Ongoing trend towards urbanization and re-densification

Cities are growing worldwide—in their spatial extent, concerning their total population and with respect to their general and actual macroeconomic significance. At the same time, cities are changing rapidly, in the technology sector, regarding mobility, energy or working environments but also in connection with ecological trends and requirements.

Table 2.4 The capitals in comparison (Beijing Statistical Yearbook; Grunewald et al. 2016)

	Beijing (2012/2014)	Berlin (2014)
Inhabitants (millions)	20.7	3.5
Administrative area (km^2)	16,400	892
Population density (Inh./km^2)	1260	3900
Green area[a] (%)	73	44
m^2 green space per capita	1	1
Nature conservation area (ha/%)	137,880/8.4	2061/2.3
Number of species (plants)	2708	2179

[a]Forests, public green, waters, agricultural land

Table 2.5 City development in Germany and China (Sources: German Federal Statistical Office and National Bureau of Statistics of China)

City size (million inhabitants)	Number of cities in Germany			Number of cities in China		
	1950	1980	2010	1953[a]	1982[b]	2010[c]
>10	–	–	–	–	–	3[d]
>1	2	3	4	9	38	187
>0.5	3	11	10	16	47	274
>0.1	48	58	63	78	137	180

[a]According to the first national population census
[b]According to the third national population census
[c]CSB (2012). China City Statistical Yearbook. Chinese Statistics Press, Beijing (National Bureau of Statistics of China)
[d]In 2014 already 6 megacities (UN 2014)

China faced a rapid urbanization process in the last three decades. It is expected that around 70% of the Chinese population will live in growing or new cities in 2025. The total urban population will then amount to 900 million people, nearly 250 million more than today (Johnson 2013).

As evidenced in Fig. 2.3, it is apparent that the greatest population density of the Chinese population is in the east. It is estimated that over 90% of the Chinese population inhabit 40% of the land mass of China. China has a long history of city development, but with the policy reform and the opening since the early 1980s, urbanization has increased rapidly (e.g., Zhao 2011; Bai et al. 2014). This manifests in large rural–urban population migrations and in the expansion of urban areas and the built environment. The scale of urbanization in China has so far been extraordinary, and there is a clear indication that it will remain so in the coming decades. Thus, the impact of the country's urban growth on biodiversity and ecosystems may surpass the extent of impacts we have witnessed across the world so far (Güneralp and Seto 2013).

China's urbanization is unique in terms of its speed, scale, and government-driven nature (Ye and Wu 2014), and the process has a low level of sustainability (Xu et al. 2016). In March 2014 the National New-type Urbanization Plan (2014–2020) has been realized (Fang 2014). One of its five main aims is for China's urban population fraction to rise by 1% per year to reach 60% by 2020 (Ye and Wu 2014). However, the observed imbalance that urban land is growing faster than population growth poses a problem (Box 2.3).

Box 2.3

Growth through state-subsidized residential construction? In China there is a risk of increasing investment ruins in the form of empty ghost towns (Fig. 2.2).

In the past decade, the urban land has grown by 78.5%, whereas the urban population has only grown by 46%. Many houses in urban areas are vacant. Cities and municipalities designated 3500 new building areas for the development of new residential areas and industrial parks, particularly in small- and medium-sized towns (Li 2016).

China's high speed urbanization boosted the economic growth and social progress, which greatly improved Chinese people's living standards and human well-being. Rapid urbanization, however, also resulted in a series of problems such as environmental pollution and ecological degradation. The current environmental pollutant emissions are undoubtedly influencing ecosystem functions and services, public health and even urban sustainability in China. China now has 690 million people living in urban areas where the environmental conditions are deteriorating, including through air pollution, especially the high PM2.5 concentrations in many megacities like Beijing, Tianjin, Nanjing, Shanghai, Hangzhou, Guangzhou, etc. (Fig. 2.1c; Sect. 3.3.3). How to control the ever-increasing environmental pollution and make cities livable for urban dwellers is a great challenge.

Fig. 2.2 The ghost city of Kangbashi, Inner Mongolia. © Wei Hou

Against this background, the general idea in the Integrated Reform Plan for Promoting Ecological Progress issued in 2015 is respecting, protecting, and staying in tune with nature. Conserving resources and protecting the environment have come to be the fundamental state policy and were given high priority (Box 2.4).

Box. 2.4 Chinese environmental and urban ecological scientists have paid much attention to the current urban environmental and ecological issues and focused on the following topics

… Urban dynamics and its impacts on urban ecosystem services, such as mitigation effects of urban green and blue infrastructure on the urban climate (Kong et al. 2014; Kuang et al. 2015; Sun and Chen 2012; Zhang et al. 2010; Zhou et al. 2014).
… Ecosystem services assessment and evaluation, urbanization impacts on urban biodiversity (Zhao et al. 2006), urban biodiversity investigation and assessment. But how to integrate these concepts and research findings into urban planning and management practices is still a challenge in China (Liu 2010).
… Ecological and environmental consequences associated with urbanization processes in China, such as urban thermal environmental mitigation (Li et al. 2011; Peng et al. 2015), air pollution reduction Sect. 3.3.3 and the impact of urban expansion on ecosystem services (Lin et al. 2013).
… Urban planning and management theories and practices to promote urban sustainable development and ecological progress.

According to the administrative divisions of China, there are three levels of cities: municipalities, prefecture-level cities, and county-level cities (Zhou et al. 2013). Municipalities and prefecture-level cities are each not a 'city' in the strictest sense of the term, but instead an administrative unit comprising, typically, both an urban core (a city in the strict sense) and surrounding rural or less urbanized areas, usually many times the size of the central, built-up core. Prefecture-level cities nearly always contain multiple counties, county-level cities, and other such subdivisions. To distinguish a prefecture-level city from its actual urban area (city in the strict sense), the term 'urban area' or 'built-up area' can be used. However, even this term encompasses large suburban regions often greater than 3000 km^2. For comparison: The administrative area of Chongqing (Table 2.6) is almost as large as the territory of Austria, and the area of Beijing is 18 times greater than the Berlin administrative area. Comparability is more likely given with respect to metropolitan regions in Germany such as the Rhine-Ruhr region with about 10 million inhabitants and 7000 km^2.

In China, the municipality is a political designation defining regions under control of a municipal government. The population of the official Chinese urban areas is listed in Table 2.6 for the five biggest cities (based on administrative area). The city of Xuzhou is added because it is an important example of structural change regarding green space implementation (Chap. 4).

We are focusing in particular on four case studies in China: Beijing, Shanghai, Chengdu and Xuzhou (Fig. 2.3, Table 2.6), because they are well distributed, they represent relevant problems, and data and case studies are available.

Germany has an established urban structure and is highly urbanized (Fig. 2.4; Table 2.7). In 2013, the share of settlement and transport area in the total area was 13.6% (StaBA 2015). 75% of the German population live in cities currently; however, two-thirds of them in small- and medium-sized cities. Only the Rhine-Ruhr metropolitan region with over 11 million inhabitants (2014) and 7110 km^2 has dimensions comparable to those of the biggest Chinese cities (Table 2.6).

Table 2.6 Population characteristics of the five biggest Chinese cities (by administrative area) and selected case study cities (Source: National Bureau of Statistics of China and Xuzhou Statistical Yearbook)

	Thousand inhabitants in 2010		Area (km^2)		Population density (inh./km^2)	
	Admin. area	Urban area	Admin. area	Urban area	Admin. area	Urban area
Chongqing	28,846	15,693	82,402	26,025	350	603
Shanghai	23,019	22,315	6340	5155	3630	4329
Beijing	19,612	18,827	16,408	12,187	1195	1545
Chengdu	14,048	7416	12,121	2064	1159	3593
Tianjin	12,938	11,091	1946	7418	1083	1495
Xuzhou	8577	1967	11,259	3037	762	647

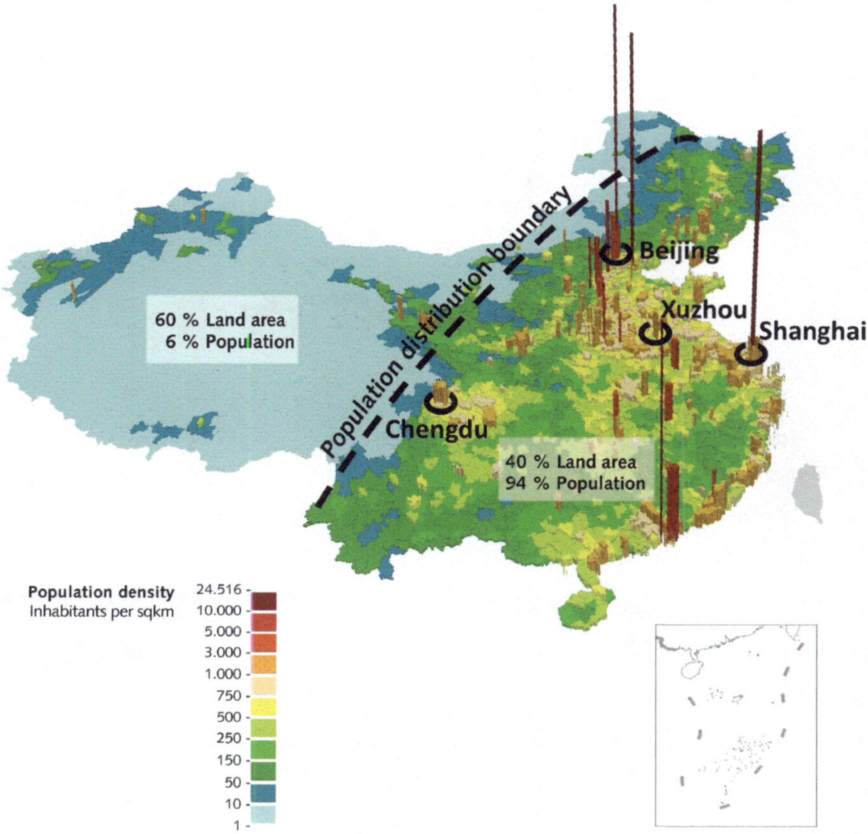

Fig. 2.3 Distribution of population density in China in 2009 and location of case study cities. © Gaodi Xie/CAS

Germany also went through a renewed process of urbanization and re-densification during the last decade (BBSR 2015). Also, in Germany more and more people want to live in the metropolitan areas, as they expect to find education and work there. Immigrants are attracted mainly by the big cities, too. Even in 'shrinking' regions, which are characterized by population decline and economic problems, new housing and transport areas continue to be developed (Haase et al. 2013).

Particularly dynamic growth is observed and predicted in the Greater Munich area, in the Berlin/Potsdam and Frankfurt/Main regions, as well as for Cologne, Hanover, and Bonn. Hamburg and the East German cities of Dresden and Leipzig are also growing, that is, the demand for housing increases, and the pressure on open spaces increases as well. These are also the German cities that were selected for the case studies (Table 2.7; Fig. 2.4; Chap. 4.3).

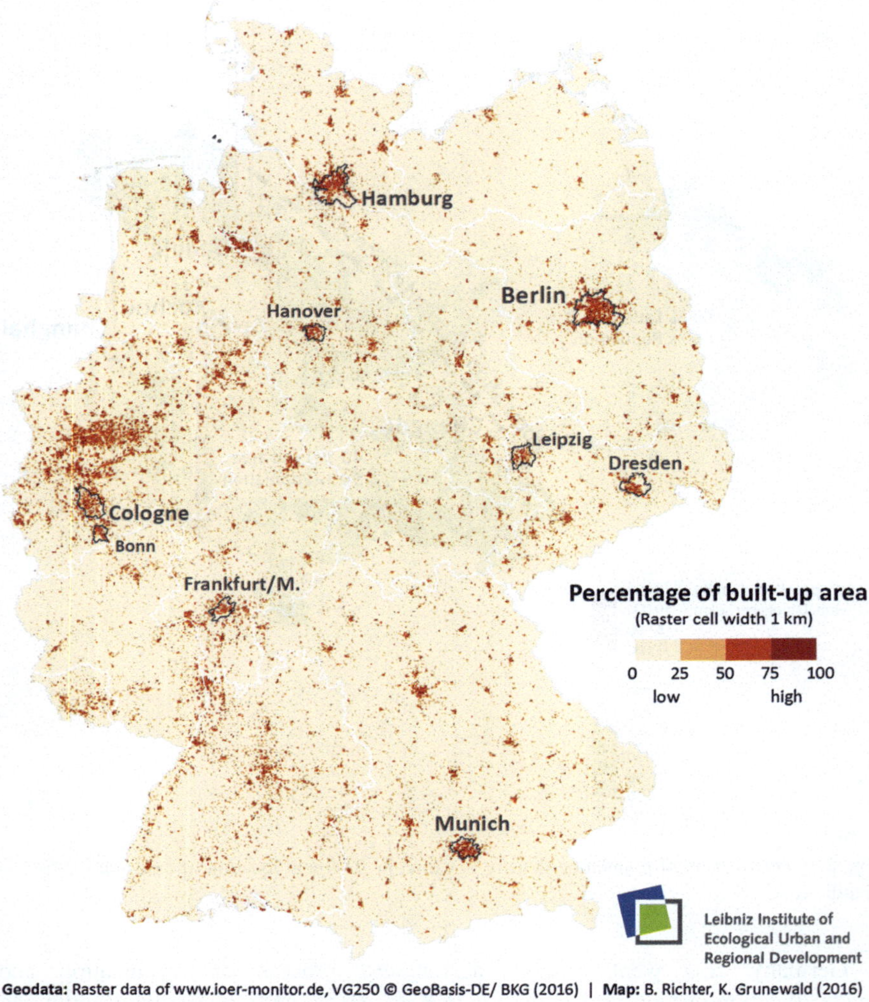

Fig. 2.4 Distribution of big cities and location of case study cities in Germany

Table 2.7 Characteristics of the five biggest German cities and selected case study cities (StaBA 2014)

	Population (rank)	Area (km^2)	Population density (inh./km^2)
Berlin	3,470,000 (1)	891.7	3890
Hamburg	1,763,000 (2)	755.3	2330
Munich	1,430,000 (3)	310.7	4600
Cologne	1,047,000 (4)	405.0	1210
Frankfurt/Main	718,000 (5)	248.3	2580
Leipzig	544,000 (11)	297.4	2430
Dresden	536,000 (12)	328.3	1630
Hanover	524,000 (13)	204.2	2570
Bonn	314,000 (19)	141.1	2230

References

Andersson E, McPhearson T, Kremer P, Gomez-Baggethun E, Haase D, Tuvendal M, Wurster D (2015) Scale and context dependence of ecosystem service providing units. Ecosyst Serv 12:157–164

Artmann M (2014) Assessment of soil sealing management responses, strategies and spatial targets towards an ecological sustainable urban land use management. Ambio 43(4):530–541

Bai X, Shi P, Liu Y (2014) Realizing China's urban dream. Nature 509:158–160

BBSR—Bundesinstitut für Bau-, Stadt- und Raumforschung (2015) Wachsen oder schrumpfen? BBSR-Analysen KOMPAKT 12/2015

Beatley T (ed) (2012) Green cities of Europe: global lessons on green urbanism. Island Press, Washington, DC

Becker H (2010) Leitbilder. In: Henckel D, Kuczkowski K, Lau P, Pahl-Weber E, Stellmacher F (eds) Planen – Bauen – Umwelt: Ein Handbuch. VS Verlag für Sozialwissenschaften/Springer Fachmedien, Wiesbaden, pp 308–312

BMU (2007) Nationale Strategie zur biologischen Vielfalt. Bundesministerium für Umwelt, Naturschutz und Reaktorsicherheit

BMUB (2015) Grünbuch Stadtgrün. Grün in der Stadt – Für eine lebenswerte Zukunft. Bundesministerium für Umwelt, Naturschutz, Bau und Reaktorsicherheit

BMUB (2017) Weißbuch Stadtgrün. Grün in der Stadt - Für eine lebenswerte Zukunft. Bundesministerium für Umwelt, Naturschutz, Bau und Reaktorsicherheit

Borgström ST, Elmqvist T, Angelstam P, Norodom-Alfsen C (2006) Scale mismatches in management of urban landscapes. Ecol Soc 11(2):16

Breuste J, Artmann M (2014) Allotment gardens contribute to urban ecosystem service: case study Salzburg, Austria. J Urban Plan Dev 141(3):A5014005. doi:10.1061/(ASCE)UP.1943-5444.0000264

Breuste J, Haase D, Elmqvist T (2013) Urban landscapes and ecosystem services. In: Wratten S, Sandhu H, Cullen R, Costanza R (eds) Ecosystem services in agricultural and urban landscapes. Wiley, Chichester, pp 83–104

Breuste J, Pauleit S, Haase D, Sauerwein M (2016) Stadtökosysteme. Funktion, Management, Entwicklung. Springer Spektrum, Berlin

Brundtland Commission (1987) Report of the World Commission on Environment and Development: Our Common Future

Bryson JM (2004) Creating and implementing your strategic plan. A workbook for public and nonprofit organizations. San Francisco

BiB—Bundesinstitut für Bevölkerungsforschung (2013) Pro-Kopf-Wohnfläche erreicht mit 45 m^2 neuen Höchstwert. Press release 9/2013 of 24 July 2013. www.bib-demografie.de/SharedDocs/Publikationen/DE/Download/Grafik_des_Monats/2013_07_pro_kopf_wohnflaeche.pdf. Accessed 28 Aug 2016

CBD—Convention on Biological Diversity (2016) https://www.cbd.int/information/parties.shtml

De Jong M, Joss S, Schraven D, Zhan C, Weijnen M (2015) Sustainable–smart–resilient–low carbon–eco–knowledge cities; making sense of a multitude of concepts promoting sustainable urbanization. J Clean Prod 109:25–38

DESA—Department of Economic and Social Affairs (2013) World economic and social survey 2013. Sustainable development challenges. Secretariat, Department of Economic and Social Affairs of the United Nations Secretariat

Die Bundesregierung (ed) (2016) Deutsche Nachhaltigkeitsstrategie – New edition 2016, as of 1 October 2016. Cabinet decision of 11 January 2017. www.deutsche-nachhaltigkeitsstrategie.de

EU—European Union (2007) Leipzig Charter on Sustainable European Cities

EU COM—European Commission (2011) Our life insurance, our natural capital: an EU biodiversity strategy to 2020. European Commission. http://eur-lex.europa.eu/legal-content/EN/TXT/?uri=celex%3A52011DC0244. Accessed 12 Sept 2016

EU COM—European Commission (2013) Green Infrastructure (GI)—Enhancing Europe's natural capital. European Commission. http://eur-lex.europa.eu/legal-content/EN/TXT/?uri=celex%3A52013DC0249. Accessed 12 Sept 2016

Fang C (2014) Strategic direction of the transformation of China's new urbanization development. Resour Environ Dev 2:18–26

Fryd O, Pauleit S, Bühler O (2011) The role of urban green space and trees in relation to climate change. CAB Rev Perspect Agric Vet Sci Nutr Nat Resour 6(053):1–18

Fürst F, Himmelbach U, Potz P (1999) Leitbilder der räumlichen Stadtentwicklung im 20. Jahrhundert – Wege zur Nachhaltigkeit. Berichte aus dem Institut für Raumplanung, vol 41. IRPUD, Dortmund

GBTimes (2015) Per capita living space is 24 square meters in Shanghai. gbtimes Beijing 20 Apr 2015. http://gbtimes.com/china/capita-living-space-24-square-meters-shanghai. Accessed 28 Aug 2016

Gill S, Handley J, Ennis R, Pauleit S (2007) Adapting cities for climate change: the role of the green infrastructure. Built Environ 33(1):115–133

GreenKeys Team (2008) GreenKeys @ your city—a guide for urban green quality. IOER Leibniz Institute of Ecological and Regional Development, Dresden. Booklet + CD-ROM

Grunewald K, Artmann M, Mathey J, Müller B, Rößler S, Syrbe R-U, Wende W, Xie X, Li J, Breuste J, Chang J, Luo P, Hu T, Kümper-Schlake L, Schröder A, Wolanska K, Xiao N, Li X, Hoppenstedt A, Xu Q, Zhu S, Spangenberg JH, Wüstemann H, Sharma K, Chen B (2016) Policy brief: towards green cities in China and Germany. Bonn. http://www.bfn.de/fileadmin/BfN/service/Dokumente/PolicyBrief_GreenCities_China_Germany_042016_en.pdf

Güneralp B, Seto KC (2013) Sub-regional assessment of China: urbanization in biodiversity hotspots. In: Elmqvist T et al. (eds) Urbanization, biodiversity and ecosystem services: challenges and opportunities: a global assessment. Springer, Dordrecht, pp 57–63

Haase D, Kabsich N, Haase A (2013) Endless urban growth? On the mismatch of population, household and urban land area growth and its effects on the urban debate. PLoS ONE 8(6): e66531

Habitat III (2016) Draft New Urban Agenda. Quito Declaration on Sustainable Cities and Human Settlements for All

Hansen R, Pauleit S (2014) From multifunctionality to multiple ecosystem services? A conceptual framework for multifunctionality in green infrastructure planning for urban areas. Ambio 43 (4):516–529

Hansen R, Frantzeskaki N, McPhearson T, Rall E, Kabisch N, Kaczorowska A, Kain J-H, Artmann M, Pauleit S (2015) The uptake of the ecosystem services concept in planning discourses of European and American cities. Ecosyst Serv 12:228–246

Harrop SR, Pritchard DJ (2011) A hard instrument goes soft: the implications of the Convention on Biological Diversity's current trajectory. Glob Environ Change 21(2):474–480

Herbst T (2014) Kommunale Biodiversitätsstrategien. Ein Werkstattbericht. Bündnis „Kommunen für biologische Vielfalt" eV, BfN – Bundesamt für Naturschutz, DUH – Deutsche Umwelthilfe eV, Radolfzell

Howard E (1898) To-morrow: a peaceful path to real reform. Swan Sonnenschein & Co., London

Huang L, Wu J, Yan L (2015) Defining and measuring urban sustainability: a review of indicators. Landscape Ecol 30:1175–1193

Hu W (2011) Research on the planning of Urban Biological Diversity Protection. Doctoral thesis

Johnson I (2013) China's great uprooting: moving 250 million into cities. New York Times, 15 June 2013

Joss S (2015) Sustainable cities. Governing for urban innovation. Planning—environment—cities. Palgrave, London

Joss S, Tomozeiu, D, Cowley R (2011) Eco-Cities—a global survey: Eco-City profiles. International Eco-Cities initiative, London. www.westminster.ac.uk/ecocities. Accessed 13 Sept 2016

Kabisch N, Qureshi S, Haase D (2015) Human-environment interactions in urban green spaces—a systematic review of contemporary issues and prospects for future research. Environ Impact Assess Rev 50:25–34

Kong F, Yin H, James P, Hutyra LR, He HS (2014) Effects of spatial pattern of greenspace on urban cooling in a large metropolitan area of eastern China. Landscape Urban Plan 128:35–47

Kowarik I, Bartz R, Brenck M (eds) (2016) Naturkapital Deutschland – TEEB DE. Ökosystemleistungen in der Stadt – Gesundheit in der Stadt – Gesundheit schützen und Lebensqualität erhöhen. Technische Universität Berlin, Helmholtz-Zentrum für Umwelt-forschung – UFZ, Leipzig, Berlin

Kuang W, Liu Y, Dou Y, Chi W, Chen G, Gao C, Yang T, Liu J, Zhang R (2015) What are hot and what are not in an urban landscape: quantifying and explaining the land surface temperature pattern in Beijing, China. Landscape Ecol 30:357–373

Kühnau C, Böhme C, Bunzel A, Böhm J, Reinke M (2016) Von der Theorie zur Umsetzung: Stadtnatur und doppelte Innenentwicklung (From theory into practice: Urban green space and dual inner development). Natur und Landschaft 7:329–335

Kuder T (2008) Leitbildprozesse in der strategischen Planung. In: Hamedinger A, Frey O, Dangschat JS, Breitfuss A (eds) Strategieorientierte Planung im kooperativen Staat. VS Verlag für Sozialwissenschaften, Wiesbaden, pp 178–192

Li J (2016) China's urbanization: problems, directions, and solutions. In: 9th Sino-German workshop, Nanjing

Li J, Song C, Cao L, Zhu F, Meng X, Wu J (2011) Impacts of landscape structure on surface urban heat islands: a case study of Shanghai, China. Remote Sens Environ 115:3249–3263

Lin T, Xue X, Shi L, Gao L (2013) Urban spatial expansion and its impacts on island ecosystem services and landscape pattern: a case study of the island city of Xiamen, Southeast China. Ocean Coast Manage 81:90–96

Liu H (2010) Biological conservation planning in urban region based on biological process: a case study in Taizhou of Zhejiang Province. Chin J Ecol 1:8–15

Liu Y, Lu S, Chen Y (2013) Spatio-temporal change of urban-rural equalized development patterns in China and its driving factors. J Urban Stud 32:320–330

Liu Z, He C, Wu J (2016) General spatiotemporal patterns of urbanization: an examination of 16 world cities. Sustainability 8(1):41. doi:10.3390/su8010041

Loibl W, Stiles R, Pauleit S, Hagen K, Gasienica B, Tötzer T, Trimmel H, Köstl M, Feilmayr W (2014) Improving open space design to cope better with urban heat island effects. GAIA 23 (1):64–66

Orlik T, Fung E (2012) In China, a move to tiny living space. Wall Street J. www.wsj.com/articles/SB10000872396390444024204578044592999502174. Accessed 28 Aug 2016

Pauleit S, Breuste JH (2011) Land use and surface cover as urban ecological indicators. In: Niemelä J (ed) Handbook of urban ecology. Oxford University Press, Oxford, pp 19–30

Pauleit S, Liu L, Ahern J, Kazmierczak A (2011) Multifunctional green infrastructure planning to promote ecological services in the city. In: Niemelä J (ed) Handbook of urban ecology. Oxford University Press, Oxford, pp 272–285

Peng J, Liu Y, Wu J, Lv H, Hu X (2015) Linking ecosystem services and landscape patterns to assess urban ecosystem health: a case study in Shenzhen City, China. Landscape Urban Plan 143:56–68

Rees WE (1992) Ecological footprints and appropriated carrying capacity: what urban economics leaves out. Environ Urbanisation 4(2):121–130

Register R (1973) Ecocity Berkeley. Building cities for a healthy future. North Atlantic Books

Rößler S (2010) Freiräume in schrumpfenden Städten. Chancen und Grenzen der Freiraumplanung im Stadtumbau. Leibniz Institute of Ecological and Regional Development: IÖR-Schriften, vol 50. Rhombos, Berlin

Schröder P (2000) Politische Strategien. Nomos Verlagsgesellschaft, Baden-Baden

Shepard W (2015) Can hundreds of new "ecocities" solve China's environmental problems? http://www.citymetric.com/skylines/can-hundreds-new-ecocities-solve-chinas-environmental-problems-1306. Accessed 3 July 2016

Sieverts T (1998) Was leisten städtebauliche Leitbilder? In: Becker H, Jessen J, Sander R (eds) Ohne Leitbild? Städtebau in Deutschland und Europa. Karl Krämer, Wüstenrot Stiftung, Stuttgart, Zürich, Ludwigsburg, pp 22–40

Sondermann M, Rößler S (2016) 9.1 Leitbilder erstellen, Orientierung bieten. In: Kowarik I, Bartz R, Brenck M (eds) Ökosystemleistungen in der Stadt: Gesundheit schützen und Lebensqualität erhöhen. Berlin, Leipzig: Technische Universität Berlin, Helmholtz-Zentrum für Umweltforschung–UFZ, pp 218–221

Spiekermann M, He Y, Yang J, Burkhardt I, Yang F, Xin Y, Pauleit S (2013) Hangzhou—fast urbanisation and high population growth. In: Nilsson K, Pauleit S, Bell S, Aalbers C, Nielsen TS (eds) Peri-urban futures: land use and sustainability. Springer, Heidelberg, pp 307–337

StaBA—Statistisches Bundesamt (2014) Statistisches Jahrbuch 2014. https://www.destatis.de/DE/Publikationen/StatistischesJahrbuch/StatistischesJahrbuch2014.pdf

StaBA – Statistisches Bundesamt (2015) Flächennutzung. Bodenfläche nach Nutzungsarten

Stiehr N (2009) Die Nationale Strategie zur biologischen Vielfalt der Bundesregierung als Diskursarena im Diskursfeld „Klimabedingte Veränderungen der Biodiversität". Institut für sozial-ökologische Forschung (ISOE), Frankfurt am Main

Sun R, Chen A, Chen L, Lü Y (2012) Cooling effects of wetlands in an urban region: the case of Beijing. Ecol Ind 20:57–64

UN—United Nations (1992) Rio Declaration on Environment and Development

UN-Habitat (United Nations Human Settlements Program) (2002) Sustainable Urbanisation: Achieving Agenda 21. UN-Habitat. Department for International Development, London, Nairobi

UN-Habitat (United Nations Human Settlements Program) (2012) http://unhabitat.org/about-us/un-habitat-at-a-glance. Accessed 12 Sept 2016

UN—United Nations (2014) World urbanization prospects: the 2014 revision, highlights (ST/ESA/SER.A/352)

UN—United Nations (2015) Sustainable development goals. http://www.un.org/sustainabledevelopment/cities/. Accessed 12 Sept 2016

UN—United Nations (2016) Draft outcome document of the United Nations Conference on Housing and Sustainable Urban Development (Habitat III)

Von Döhren P, Haase D (2015) Ecosystem disservices research: a review of the state of the art with a focus on cities. Ecol Ind 52:490–497

Wang H, Qureshi S, Qureshi BA, Qiu J, Friedman CR, Breuste JH, Wang X (2016) A multivariate analysis integrating ecological, socioeconomic and physical characteristics to investigate urban forest cover and plant diversity in Beijing, China. Ecol Ind 60:921–929

WBGU—Wissenschaftlicher Beirat der Bundesregierung Globale Umweltveränderungen (2016) Der Umzug der Menschheit: Die transformative Kraft der Städte. Wissenschaftlicher Beirat der Bundesregierung Globale Umweltveränderungen, Berlin

Xu C, Wang S, Zhou L, Wang L, Liu W (2016) A comprehensive quantitative evaluation of new sustainable urbanization level in 20 Chinese urban agglomerations. Sustainability 8(2):91. doi:10.3390/su8020091

Ye L, Wu A (2014) Urbanization, land development, and land financing: evidence from Chinese cities. J Urban Aff 36(s1):354–368

Yu D, Jiang Y, Kang M, Tian Y, Duan J (2011) Integrated Urban Land-Use Planning Based on Improving Ecosystem Service: Panyu Case, in a Typical Development Area in China. J Urban Plann Dev 137(4):448–458

Zhang K, Wang R, Shen C, Da L (2010) Temporal and spatial characteristics of the urban heat island during rapid urbanization in Shanghai, China. Environ Monit Assess 169:101–112

Zhao J (2011) Towards sustainable cities in China. Analysis and assessment of some Chinese Cities in 2008. SpringerBriefs in Environmental Sciences. Springer, New York

Zhao SQ, Da LJ, Tang ZY, Fang HJ, Song K, Fang JY (2006) Ecological consequences of rapid urban expansion: Shanghai, China. Front Ecol Environ 4(7):341–346

Zhou S, Dai J, Bu J (2013) City size distribution in China 1949 to 2010 and the impacts of government policies. Cities 32(1):51–57

Zhou W, Qian Y, Li X, Li W, Han L (2014) Relationships between land cover and the surface urban heat island: seasonal variability and effects of spatial and thematic resolution of land cover data on predicting land surface temperatures. Landscape Ecol 29:153–167

Chapter 3
The Multiple Benefits of Urban Green—Ecosystem Services Assessment

Karsten Grunewald, Gaodi Xie and Henry Wüstemann

The elements of the green infrastructure in the urban space represent habitats for numerous species and provide a high variety of ecosystem services. A clarification of the benefits urban residents gain from urban green spaces as well as control of urban green quality at the city and federal level are necessary to derive and pinpoint urban planning and nature conservation strategies in this context. In this light, due to its integrated approach the concept of ecosystem services (ES) is gaining increased popularity in the global environmental debate.

In the context of our Green Cities Study, opportunities will be identified how human benefits (health effects, Sect. 3.1) and biodiversity (Sect. 3.2) can be promoted through the strengthening of urban green space and the evaluation of ES at the same time in urban areas. The selection of evidence-based research is based on the current state of knowledge internationally and in Germany and China.

Since the book focuses on issues relevant for urban areas, the attention is on direct and locally generated ES. To demonstrate the benefits, quantification in nonmonetary and monetary terms is extremely helpful. We are particularly focusing on the typological level and calculating ES of forests, parks, green spaces, water bodies, etc. Green space in cities exists in a broad variety of types spanning from high-maintenance urban parks to natural areas and buffer space between noisy infrastructure and other land uses. From such a degree of heterogeneity in the type

K. Grunewald (✉)
Leibniz Institute of Ecological Urban and Regional Development, Dresden, Germany
e-mail: k.grunewald@ioer.de

G. Xie
Institute of Geographic Sciences and Natural Resources Research, Chinese Academy of Sciences (CAS), Beijing, China
e-mail: xiegd@igsnrr.ac.cn

H. Wüstemann
Technische Universität Berlin, Berlin, Germany
e-mail: henry.wuestemann@tu-berlin.de

K. Grunewald et al. (eds.), *Towards Green Cities*, Cities and Nature,
DOI 10.1007/978-3-319-58223-8_3

of green spaces it follows that the benefits (and costs) generated by different types of green space provision vary greatly (Panduro and Veie 2013). Location, structure, composition, and spatial configuration of urban green spaces influence the ecological qualities and services.

Examples illustrate what significance and relevance ES have. The examples relate mostly to the specific situation in China and Germany. There was a need to keep sections as short as possible, so that individual problems could not be presented explicitly and in details.

A focal point of this chapter will be placed on regulating ES (Sect. 3.3), which contribute significantly to environmental quality in cities. Accessibility of public green spaces as cultural ES is the focus of Sect. 3.4. Ecosystems in cities deliver services for the residents regarding the experience of nature, recreational activities, relaxation, outdoor dancing and sports, and aesthetics. In particular, they contribute to the psychological and physical regeneration of the population and are therefore in demand. In principle, urban nature also has adverse effects on individuals or on society, so-called disservices (e.g., von Döhren and Haase 2015). For reasons of space, however, it is not possible to consider these further in this study.

Provisioning services aspects are examined within examples of urban agriculture or food and water provision (Sect. 3.5), but not shown in greater depth, because for instance the food produced within the city can only supplement the supply of residents. Supporting ES such as soil formation and nutrient cycling are not systematically addressed, since they have no direct connection to human well-being.

Comparing ES values with potential technical alternatives clearly shows that elements of green infrastructure are not only ecologically and socially desirable, but often economically beneficial, providing services at lower costs than technical solutions or healthcare (Sect. 3.6). On the other hand, provision of green space in a dense urban environment is costly. The rent from alternative land-use for areas allocated to green space would be high.

If urban green space is so important and has so many benefits, the question ultimately arises how green our cities really are. This will be addressed in Sect. 3.7. How much "green" is enough? The establishment of target values is difficult. In addition to the proportion and volume of green space, spatial arrangement, accessibility, and other qualitative aspects play a key a role.

3.1 Well-Being and Health Effects of Urban Green Space

Henry Wüstemann and Yi Zhang

A considerable number of studies investigating well-being and health benefits exist at the international level. Literature originating from psychology shows that urban green has positive effects on mental health by reducing stress (Grahn and Stigsdotter 2003; Swanwick et al. 2003; Stigsdotter et al. 2010) and mental distress (Annerstedt et al. 2012; Richardson et al. 2013; Sturm and Cohen 2014). Moreover, exposure to green significantly reduces rates of anxiety and depression (de Vries et al. 2003; Maas et al. 2009) and can raise positive emotions (Ulrich et al. 1991; Knecht 2004; Bowler et al. 2010a; Coon et al. 2011). Interestingly, even short-term exposure to green seems to improve cognitive functioning and mood (Ulrich et al. 1991; Hartig et al. 2003; Abkar et al. 2010). A second subset of literature originating from medicine shows that urban green improves physical health (de Vries et al. 2003; Maas et al. 2006; Richardson et al. 2013) and longevity (Takano et al. 2002). More recently a new stream of literature provides evidence that urban green space positively influences life satisfaction in general (Smyth et al. 2008; Alcock et al. 2014; Krekel et al. 2016).

Besides restorative effects in mental health and increasing physical activities as discussed above, the mechanisms through which green space benefits health also include improving air quality, including through removing ozone (Calfapietra et al. 2016) and storing carbon dioxide (Mcpherson 1997; Chen 2015), buffering anthropogenic noise (Tyagi et al. 2013), decreasing the urban heat island effect (Sect. 3.3.1), improving the immune system through microbial input from the environment to drive immunoregulation (Rook 2013), and fostering social cohesion (Newton 2007).

Although the causal relationships of the positive effects of green have not been sufficiently clarified, an increase in physical activities (Kaczynski and Henderson 2007; Maas et al. 2008), a rise in social activities (Kuo et al. 1998), as well as social cohesion and identity (Newton 2007) represent potential transmission channels for the positive impact of green on well-being and health.

Measuring benefits of urban green for life satisfaction and health

Studies investigating well-being and health effects of urban green space apply a high variety of well-being and health-related indicators, including *Satisfaction With Life* (SWL), *Perceived General Health* (PGH), *Health-Related Quality of Life* (HRQOL), and various stress-related indicators such as cortisol levels (e.g., Maas et al. 2006; Honold et al. 2015; Krekel et al. 2015, 2016).

Perceived General Health (PGH) is self-rated by replying to the question "In general, would you say that your health is…" (Rütten 2001). The respondent can choose between the five categories "very good," "good," "neither good nor poor," "poor," and "very poor." These five categories are often dichotomised with "neither good nor poor" (0) as the cut-off point (Maas et al. 2006).

One of the most popular instruments is Health-Related Quality of Life (HRQOL), which refers to functioning and well-being in physical and mental dimensions of life (Farivar et al. 2007; Wüstemann et al. 2017). The physical and mental health summary scores SF-36 and SF-12 are the most frequently used multi-item HRQOL instruments (Farivar et al. 2007) and have been applied to population studies (e.g., Bullinger et al. 1998; Fukuhara et al. 1998) as well as clinical trials (Hlatky et al. 2002). The SF-36 is a multi-dimensional instrument that measures the self-assessed overall health status and health-related life satisfaction (Ellert and Kurt 2004). The SF-36 consists of eight multi-item scales assessing physical function, bodily pain, general health, vitality, social functioning, role limitations, and emotional well-being (Farivar et al. 2007). The SF-12—a 12-item subset of the SF-36—incorporates the two summary measures "physical and mental component summary scores" (Ware et al. 2000; Farivar et al. 2007).

The indicator "Satisfaction with Life" is usually obtained from an eleven-point single-item Likert scale which asks "How satisfied are you with your life, all things considered?" The higher the score on the Likert scale, the higher the level of life satisfaction. Besides "Satisfaction with Life," the indicator "Job Satisfaction" covers another area of life that is heavily investigated in relation to life satisfaction (Judge and Watanabe 1993). A large number of studies have investigated the impact of socioeconomic variables on life satisfaction. Findings of these studies indicate that variables such as income and employment have a significant impact on life satisfaction (e.g., Ferrer-i-Carbonell 2005; Kaplan et al. 2008; McKee-Ryan et al. 2005). This is especially important as these findings can be used as controls in studies investigating the impact of urban green on life satisfaction.

Studies investigating well-being and health effects of urban green in China and Germany

While a large body of literature exists at the international level, studies investigating the impact of urban green on life satisfaction and health are relatively scarce in China and Germany.

Smyth et al. (2008) investigate the relationship between environmental factors such as atmospheric pollution, traffic congestion, and access to parkland and residential well-being in urban China by looking at 30 cities. The study provides evidence that residents living in cities with greater access to urban parks report higher levels of well-being. Using cross-section data from a web survey in 2012 and cross-section data from the European Urban Atlas (EUA), Bertram and Rehdanz (2015) found that green has positive effects on the life satisfaction of residents in Berlin. According to this study, the amount of green space in a 1 km buffer that leads to the largest positive effect on life satisfaction is 35 ha or 11% of the buffer area, whereas 75% of the respondents have less green space available. However, the study has some limitations. First, the study investigates the effects of urban green on residential well-being in the city of Berlin only. Krekel et al. (2016) argue that Berlin might be a special case in the sense that it has a higher share of green when compared to other major German cities. Moreover, Bertram and Rehdanz (2015) used survey data with a relatively small sample size and imprecise measures of the geographical locations of the places of residence of individuals. The only nationwide study in Germany investigating the impact of urban land use on residential well-being and mental and physical health in 32 major German cities with more than 100,000 inhabitants has been carried out by Krekel et al. (2016). Besides *Satisfaction with Life* (SWL), Krekel et al. also apply numerous variables measuring Health-Related Quality of Life (HRQOL) to this study. In addition, the study also examines the impact of access to urban green (self-assessed) on particular disease patterns such as back complaints, diabetes, joint disease, and sleep disorder. The study proves the relevance of urban green for residential well-being, including the finding that residents who live closer to urban green or are surrounded by higher amounts of green show higher life satisfaction levels. Moreover, the study found that urban green, forests, and waters positively influence mental and physical health, including social functioning, vitality, bodily pain, and overall mental health. The study further found that residents with improved access to green report a lower Body Mass Index (BMI). Moreover, exposure to abandoned areas adversely affects residential well-being, including life satisfaction, mental and physical health, and the Body Mass Index (BMI). The study also shows that residents who live closer to green suffer significantly less often from particular diseases such as diabetes, sleep disorder, and joint disease (Table 3.1).

Honold et al. (2015) examine the cross-sectional relationships between two kinds of urban nature (neighborhood vegetation visible from the home, use of public green spaces) and health outcomes including Satisfaction with Life (SWL), Perceived General Health (PGH), and 2-month hair cortisol levels in the city of Berlin. The study uses a relatively small sample size of only 32 participants. However, the study shows that participants whose homes had views of large amounts of diverse kinds of vegetation and who regularly used a vegetated trail along a canal had lower cortisol levels and reported significantly higher life satisfaction levels.

Conclusion

The studies investigating the well-being and health effects of urban green space in China and Germany presented here show how important green space is for human well-being in urban areas. The studies further provided helpful information for policy and planning on the optimal amount of green space provision for well-being and health in close vicinity of the households. Moreover, green space has the potential to reduce socioeconomic health inequalities in urban areas. Considering the construction and maintenance costs of additional green space, the well-being benefits of green space would outweigh the costs.

Although the studies on well-being and health effects of urban green represent an important step toward an understanding of the multiple benefits of green space, there is a lot of room for further research. Specifically, future investigations should be directed toward establishing the causality of these effects in order to identify possible transmission channels of the positive effects of urban green space on residential well-being and health. Moreover, further research should focus on the quality of urban green space and health. The growing availability of geocoded urban land-use data potentially provides new opportunities in this field of research.

The effect of health and well-being would depend on the availability, accessibility, and utilization of urban green space. The causal relationships between health effect and urban green have not been completely established. However, we recognize that urban green can have multiple causal links with human health and well-being. The evidence for specific health effects, such as improving the human immune system and improving air quality, has been accumulated through experiments (Kuo 2015). In future work, our understanding of the causal relationships can be enhanced by transdisciplinary cooperation and modeling.

Table 3.1 Overview of studies investigating the impact of urban green on well-being and health in China and Germany

Study	Purpose	Indicators	Main findings
Bertram and Rehdanz (2015)	Impact of urban green on residential well-being in Berlin (Germany)	Satisfaction with Life (SWL)	Findings show that an amount of green space of 35 ha (or 11%) in a 1 km buffer around households leads to the largest positive effect on SWL. The study further shows that 75% of the respondents have less green space available in their neighborhood
Xu and Wu (2015)	Relationship between residential green environment and life satisfaction, physical and mental health in Qingdao (China)	Life satisfaction, physical, and mental health	Positive correlation between residential green environment and life satisfaction
Smyht et al. (2008)	Relationship between environmental factors (e.g., access to parkland) in 30 cities in China	Satisfaction with Life (SWL)	Residents living in cities with greater access to urban parks report higher life satisfaction levels
Krekel et al. (2016)	Impact of urban land use on residential well-being in major German cities	Satisfaction with Life (SWL)	Positive impact of forests, urban parks, and waters on residential well-being. Negative effects of abandoned areas. Moreover, especially older residents benefit from the well-being effects of urban green. The SWL benefits of urban green space would outweigh the costs for construction and maintenance of additional green space
Krekel et al. (2015)	Impact of urban land use on well-being and mental and physical health	Health-Related Quality of Life (HRQOL)	Positive relationship between forests, urban parks and waters and HRQOL and negative impact of abandoned areas on HRQOL
Honold et al. (2015)	Relationship between urban nature and health outcomes in the city of Berlin (Germany)	Satisfaction with Life (SWL), Perceived General Health (PGH), cortisol level	Views of vegetation and use of vegetated trails result in lower cortisol levels and higher life satisfaction

3.2 The Role of Biodiversity in a City

Olaf Bastian and Nengwen Xiao

Plants, animals, and microorganisms including fungi are the basis of all ecosystems and the services they provide. Biodiversity and human well-being "…are inextricably linked" (MEA 2005, Sect. 1.2), also in cities. The question is whether it is actually the case that a high biodiversity is necessary for the provision of ES, also in cities? To date, a significant number of studies and meta-analyses of the scientific literature have shown the role of biodiversity for the supply of ES (e.g., Harrison et al. 2014). They were able to reveal that there are many strong relationships, primarily for several regulating and cultural services. Some studies show that some ecosystems need only few species to deliver what we want from them (Haase et al. 2014).

It may come as a surprise that at least regarding vascular plants and almost all animal groups, cities are characterized by great species richness; they can be important sites of local and regional biodiversity. Within the boundaries of cities the richness of land-use types and intensities of use creates a great array of different habitats, microhabitats and highly varied habitat mosaic configurations. To this is added the intentional or unintentional introduction of a great number of non-native animal and plant species.

Cities host a wide range of species, which are adapted in different ways to human settlements. There are species which occur predominantly or exclusively in cities (so-called urbanophile species, fully synanthropic species or urban exploiters), species which occur both in urban areas and in the wider landscape (urbanoneutral, casually synanthropic species, urban adapters), but also species which avoid urban spaces and habitats (urbanophobe, nonsynanthropic species, urban avoiders) (BfN 2009).

Ecological characteristics of cities that are relevant for urban biodiversity comprise the existence of:

- Dry and heat islands
- Small structures, small-scale spatial patterns of site conditions and land use
- Wilderness on fallow land
- Frequently disturbed biotopes
- High share of thermophilic and non-native species
- Refuge and substitute habitats, stepping stones

For the differentiation of nature in cities, the degrees of hemeroby, synanthropy, and naturalness are suitable indicators (Table 3.2). The original site conditions (the city's natural history) are also important; they are traced especially in less urbanized areas (Box 3.1). In this way, a high diversity in biotopes and species may occur, which may be higher than in intensively used agricultural areas. The increasing urbanization and concentration of construction, however, may cause the decline of biodiversity: Many existing species are replaced by a small number of widespread

Table 3.2 Hemeroby of vegetation (modified after Blume and Sukopp 1976 and Bastian and Schreiber 1994)

Hemeroby	Anthropogenic vegetation changes	Ecosystem types (examples)
Ahemerobic	No	Remnants of pristine nature
Oligohemerobic	Rather low/few	Near-natural forests, bogs
Mesohemerobic	Medium	Semi-natural or more artificial forests, heaths, dry and rough meadows, extensively used meadows and pastures
β-euhemerobic	Strong	Intensively used grassland and forests, perennial ruderal vegetation, segetal communities (on arable fields)
α-euhemerobic	Very strong	Special cultures (fruit orchards, vineyards, ornamental lawns), annual pioneer and ruderal vegetation
Polyhemerobic	Extremely	Waste deposits, spoil heaps, paved paths, graveled railroad tracks
Metahemerobic	Vegetation totally removed	Poisoned ecosystems, totally built-up/sealed areas (buildings, tar surface)

and competitive species. This process with a few winners and many losers is termed biotic homogenization (BfN 2009).

Box 3.1 Four basic forms of nature found in cities (Kowarik 2005):

- Nature 1: Remnants of primary "original" natural landscapes (woods, wetlands, rocky areas, etc.)
- Nature 2: Agricultural cultural landscape arose through tradition of modern agriculture and forestry (fields, meadows, meadows with mixed fruit trees, heathland, dry grassland, forests, etc.)
- Nature 3: Horticulturally designed symbolic nature (parks, gardens, street trees, flower tubs, etc.)
- Nature 4: Specific urban–industrial nature (scraggy wall vegetation, vegetation on sidewalks, spontaneous vegetation on gaps between buildings, and on abandoned areas, etc.)

The conservation of biodiversity on all levels is a crucial goal of international environmental policy. At least according to German legislation, this goal applies also to urban areas, which are also an object of the National Strategy on Biological Diversity.

Protecting biodiversity in urban areas must include all categories of nature, from the remnants of natural landscapes (e.g., near-natural forests) to cultural landscape (orchards), urban–industrial areas including rural settlements and landscaped areas. Settlements can host important substitute habitats for various threatened species of natural and cultural landscapes (Müller and Abendroth 2007).

Within the administrative borders of cities even nature reserves, landscape protection areas, and natural monuments can be situated, in the countries of the European Union also sites of the EU-wide Natura 2000 network. The latter represents one of the world's most ambitious approaches to biodiversity conservation. It is designed to help achieve the goal of the European Union "to halt the loss of biodiversity and the degradation of ES in the EU by 2020, and restore them as far as feasible, while stepping up the EU contribution to averting global biodiversity loss" (COM 2011). Natura 2000 comprises two basic categories: (1) especially and strictly protected species, and (2) protected areas with FFH sites and Special Protection Areas (SPA, for birds).

The main tasks of nature conservation in the city are not restricted to the protection of threatened plants and animals, as is normally assumed in the open countryside. Its focus lies to a great extent on biodiversity of the urban–industrial and the landscaped nature, which is specific for urban settlements. In cities, the preservation of living beings and communities due to the importance of residents' direct contact with natural environmental elements is particularly emphasized (Breuste in Bastian and Steinhardt 2002).

The "green" in cities, the so-called "green infrastructure," has a significant role in delivering ES, e.g., in absorbing pollutants from the air, sequestering C, contributing to rainwater infiltration, providing shade, cooling the air through tree transpiration, and reducing energy consumption in summer and the urban heat island (UHI) effect. By wise choice of species and design of spaces, and by increasing green surfaces (urban greening) at the ground surface, on roofs, and on vertical walls, these benefits can be increased.

From a better connection of biodiversity and ES in the public awareness, a great opportunity arises to reach two goals faster and more sustainably: (1) the maintenance of biodiversity in urban areas, and (2) to keep and enhance ES in urban areas in order to raise their attractiveness and life quality (BfN 2009).

Example 3.1 Biodiversity and its conservation in the city of Dresden
Dresden (ca. 525,000 inhabitants) is the capital city of the German federal state of Saxony. The territory stretches over several physical landscapes. This finds expression in the vegetation and the fauna, which has adapted to it, but buildings, surface sealing, and intensive land use have changed the original conditions drastically. Nevertheless, the Elbe river with its broad floodplain mostly covered by semi-natural meadows crosses the entire city from southeast to northwest. The inner city is greened by parks and avenues but also by many small habitats of partly rare and threatened plants and animals (Fig. 3.1).

Spaces and structures in Dresden with high significance for species and biotopes are as follows:

- Large-scale rather natural landscapes (e.g., a large forest within the administrative borders of the city, and the Elbe river floodplain with its tributary valleys)
- Biotopes of the cultural landscape with diverse habitats for plants and animals (e.g., extensively used grassland, orchards, woods)

Fig. 3.1 Valuable green spaces in the city of Dresden: **a** richly blossoming ruderal vegetation on brownfields, **b** historical park, **c** orchard meadow as an example of semi-natural biotopes, **d** Elbe river valley with the Pillnitz castle, floodplain meadows, and a protected river island (*right*); Strictly protected animal species in the city of Dresden: **e** Sand lizard (*Lacerta agilis*) (© **a–e** O. Bastian), **f** Juvenile Common Kestrels (*Falco tinnunculus*) in the nesting box at a building (© M. Lehnert), and **g** Lesser Horseshoe bat (*Rhinolophus hipposideros*) (© R.+E. Francke)

- Habitats and migration corridors of protected animal species (e.g., beaver, otter, bats, white stork, amphibians)
- Habitats of protected building-inhabiting animal species (e.g., bats, some bird species)
- Inner-city empty spaces and carefully managed parks as important habitats particularly for synanthropic species and as stepping stones in biotope compound systems

According to German (and European) legislation all native bird species (except, e.g., feral domestic pigeons) and bats, including their habitats, are protected. This means that it is not allowed to pursue or even kill them. In order to maintain or increase the populations of birds and bats in or at buildings, more than 16,000 nesting aids for birds have been installed in Dresden since 1997, among them 15,000 breeding stones for common swifts (*Apus apus*, more than half of them occupied, with rising tendency), and 2000 habitat stones for bats at buildings for nursery roosts. For example, the common kestrel (*Falco tinnunculus*) has 280 nesting places in Dresden (160 occupied in 2015).

Development planning is obligated to care for protected species. Such plans have to specify avoidance measures (planning alternatives, construction field preparation outside the breeding season) and compensation. Measures to ensure Continuous Ecological Functionality (CEF) are obligatory and must be implemented completely before the construction works begin, e.g., creation of substitute habitats for birds, sand lizards, rare beetles, and butterflies.

Example 3.2 Biodiversity and its conservation in the city of Beijing
Beijing has a rather abundant vascular plant diversity (2276 species), including 207 species of conservation concern such as endemic, threatened, and protected species. The exurban region has not only the highest species diversity (1998 species), but also the most species of conservation concern (194 species). The urban region possesses the most alien species in terms of both absolute number and proportional representation, while the suburban region has the lowest species diversity (1026 species). Some problems, such as wetland shrinkage and biological invasions, are common in the entire Beijing Municipality (Wang et al. 2007). However, primary threats to biodiversity differ in the three functional regions. The urban and suburban regions mainly suffer from habitat loss and fragmentation due to urban sprawl, while the exurban region faces serious ecosystem degradation from increasing disturbance from both local and urban people. Based on our investigation, we put forward conservation strategies for the three regions: improving the structure and ecological function of green spaces in the urban region, preserving as many remnant natural habitats as possible in the suburban region, and restricting rural tourism and establishing a biosphere reserve in the exurban region. In addition, improving public education and orienting it more toward social aspects of conservation practice is strongly recommended (Wang et al. 2007).

An investigation of urban plants in Beijing shows a high homogeneity (Xiao 2015). There are spatial differences in the plant distribution in Beijing's built-up area. The number of plant species is lowest within the second ring road and highest outside the fifth ring road (Fig. 3.2). However, there is little difference between the ring roads in the distribution of plant species. There are 156 species of plants that are widely distributed across all the five ring roads, accounting for 29.1% of all species. The evenness index (a measure of biodiversity which quantifies how equal the community is numerically) was similar among ring roads. The main causes of the homogenization of plants are the high intensity of human disturbance, the invasion of alien species, and the reduction of native species. Green plant cultivation and configuration are affected by human activities; especially the trees and

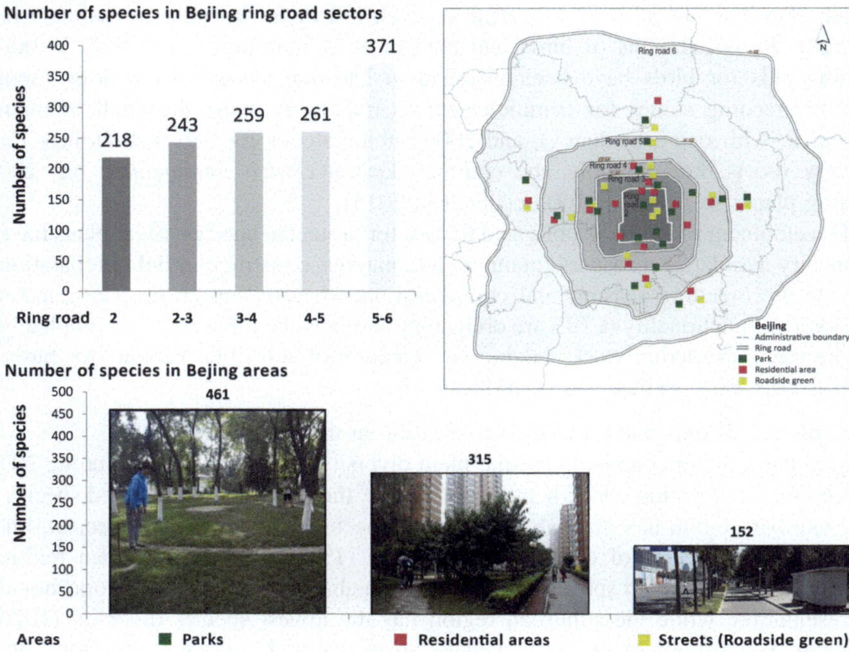

Fig. 3.2 Number of species investigated in Beijing (Xiao 2015)

shrubs on the streets are relatively poor in species, making the city plants more homogeneous.

The distribution of plants differs between habitats. Of 57 streets and 114 parks and residential greenland plots investigated in Beijing, park greenlands have the highest plant diversity, followed by residential greenlands, and street greenlands have the lowest diversity, with only 152 species (Fig. 3.2).

Beijing also has a rather abundant bird diversity (435 species), accounting for 3.76% of all Chinese species, including 34 species of "Key Bird Protection Level One" of Beijing (Xiao 2015). Urbanization has great influence on the distribution of birds. The Shannon–Wiener diversity index and the Pielou index exhibit a gradually increasing trend from the 2nd–3rd ring road to the 6th ring road. The diversity and evenness of birds decreases with increasing urbanization. However, within the second ring, the diversity index and evenness index were higher than between the 2nd and 5th ring roads. This may be due to the long history and less intense development in recent decades within the second ring road.

There are also differences in the bird distribution between habitats. There is high similarity between common commercial, farmland, residential, and industrial areas. Forests and wetlands have high similarity. Forests, wetlands, and residential areas have the largest number of birds; forests, wetlands, and farmland have a higher Shannon–Wiener diversity index and Pielou index, whereas industrial, commercial, and residential areas have lower diversity. There are a large number of birds in the residential area, but the diversity is low, mainly due to a large number of *Passer montanus* (Fig. 3.2).

3.3 Regulating Services

3.3.1 Regulation of Microclimate by Urban Green Spaces

Junxiang Li and Juliane Mathey

Urban climate associated with urbanization

Urban areas suffer from special climatic conditions: The phenomenon of the urban heat island (UHI) is characterized by dryness, heat, and lower wind strengths compared to the rural surroundings (Chou and Zhang 1982; Arnfield 2003) and associated with the urban dry island, urban moisture island, urban turbidity island, and rainfall island effects (Zhou 1990; Zhou and Wang 1996; Kuttler et al. 2007). The UHI can be measured by air temperature through traditional meteorological observation and by urban land surface temperature derived from satellite remotely sensed thermal band data (Voogt 2002). The UHI is influenced by four factors: the dark and dense materials used for paving and building, the three-dimensional structure of urban buildings, vegetation abundance, and the additional anthropogenic heat sources (Larsen 2015). The very dark, dense surface materials and the vertical structure of buildings absorb solar radiation, restrict air circulation, and prevent the penetration of the rural cold air streams in cities. Moreover, due to the high building densities, the turbulent exchanges between the urban land surface and the atmosphere are reduced, causing a heat trapping effect in streets that can be detrimental for human health during extreme heat events. The urban climate is also affected by the vegetation structure, the urban landscape structure, and the pattern of urban development (Li et al. 2011; Lehmann et al. 2014). Urban vegetation regulates urban microclimate by shading and evapotranspiration. The key ecological characteristics of cities are the structure and alignment of buildings, the proportion of sealed surface and the quality and structure of green space as well as the specifics of land use. The location, spatial distribution, and structural pattern of land-use types, and the resulting proportions of open and green spaces, will determine the resulting ecosystem services (Li et al. 2011, 2012; Mathey et al. 2011; Zhou et al. 2013; Lehmann et al. 2014).

More than half of the world population has been living in cities since 2007, and the urban population was projected to increase by 95% in the next few decades, resulting in further urbanization of the world (UNPD 2012). Urbanization increases the air and surface temperature in urban areas, which consequently intensifies the urban heat island effect (Oke 1973; Streutker 2003). Previous studies demonstrated

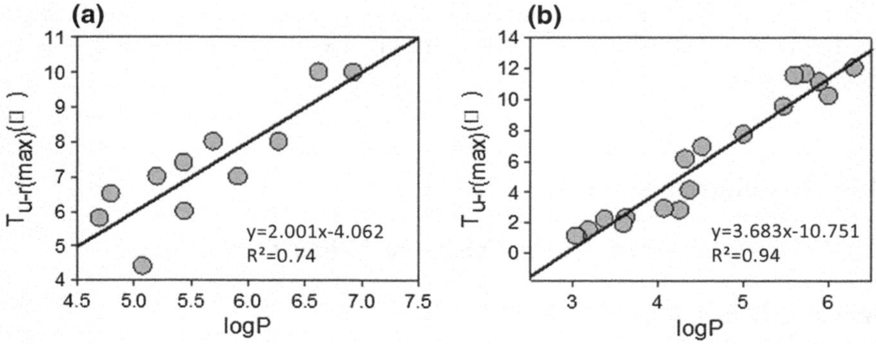

Fig. 3.3 The relationship between UHI and urban population size. **a** For cities in Europe, and **b** for cities in North America (Redrawn from Oke 1973)

that the UHI increased linearly with the logarithm of the city's size, measured by the city's population, as illustrated in Fig. 3.3.

It is expected that climate change will exacerbate these specific urban climatic conditions, further lowering the quality of life in urban areas. Cities will be exposed to global warming in its respective regional specificities. The Intergovernmental Panel on Climate Change (IPCC) in its fifth progress report estimates a further increase in extreme weather conditions. Recent climate projections show rising average temperatures in urban areas. In particular, heat waves with higher frequencies and duration will greatly undermine the quality of life in cities (IPCC 2013; Li and Bou-Zeid 2013; TEEB DE 2016). Urbanization and global warming together exacerbate the urban thermal environment. Higher temperature and the associated risks from heat wave events in urban areas will undoubtedly further influence urban ecosystems and public health (e.g., Patz et al. 2005). Therefore, how to adapt to the adverse consequences of increased temperature resulting from urbanization and climate change is a great challenge.

Heat stress and human health

Under the current trends of global warming, summer heat waves will occur more frequently (Easterling et al. 2000; Frich et al. 2002) and the risks of extreme summer heat waves were estimated to double (Schär and Jendritzky 2004; Schär et al. 2004). In summer, direct incident solar radiation, high air temperature, and humidity as well as low wind speed can cause heat load and extreme heat stress for humans (Jendritzky et al. 2009). The UHI and extreme heat events can increase the heat-related morbidity and mortality. For instance, there were at least 700 excess deaths during the record-setting heat wave in Chicago in July, 1995 (Semenza et al. 1996). The 2003 summer heat wave during August caused heat-related deaths amounting to between 22,000 and 35,000 across all of Europe. The mortality rate increased by 54% in France, and 900–1300 excess deaths were recorded in

Table 3.3 The excess mortality of urban inhabitants caused by two urban heat wave periods (HW1 and HW2) in Shanghai in 2003 (from Wang 2013)

	HW1	CK1	Excess Deaths	RR (95% CI)	HW2	CK2	Excess Deaths	RR (95% CI)
Total Deaths	4815	4212	603	1.14 (1.08,1.22)*	3428	3195	233	1.07 (1.02,1.14)*
Deaths due to Cardiovas	1627	1351	276	1.20 (1.07,1.38)*	1127	1037	90	1.09 (0.98,1.22)
Deaths due to Respiratory	624	577	47	1.08 (0.87,1.43)	488	449	39	1.09 (0.89,1.40)

CK is the control. RR is the ratio of HW to CK. $^{*}p < 0.05$

Baden-Württemberg, Germany during two weeks (Schär and Jendritzky 2004). During the summer in Shanghai in the year of 2003, three heat waves occurred from July to September 6, lasting 33 days in total with maximum air temperatures over 35 °C. A total of 836 excess heat-related deaths were estimated during the heat waves (see Table 3.3, the last two heat waves were merged into one since there were only two days between them).

In large cities, heat islands with "tropical nights" above 20 °C make it difficult to have the necessary recreation from the heat stress of the day. Sleep can be affected negatively, which may pose health hazards (Höppe 1999). Vulnerable people like elderly people, sick persons, and young infants (toddlers) are thus exposed to higher health hazards during heat waves in cities (Burkart et al. 2013; Scherber et al. 2013). Due to demographic change in Germany, the number of elderly people will increase more and more, so that climate-regulating ecosystem services will gain in importance in future. However, the heat islands are not spread equally over urban areas. Mappings exist from numerous cities, which show zones of human heat stress, mostly resulting from too dense building and from lack of green structures (Scherber et al. 2013; Krüger et al. 2014). The map in Fig. 3.4 allows the extent to which people of all age groups are affected by urban heat islands to be read off for the city center of Dresden, Germany and for Detroit, Michigan, USA.

Microclimate regulation by urban green spaces

By contributing to lower temperature, higher humidity and improved air circulation, urban green spaces have a positive influence on the health and well-being of residents (Sect. 3.1). Urban green spaces affect climate elements in manifold ways (Table 3.4).

Cooling effects of urban green spaces

Urban green spaces have cooling effects on the urban thermal environment through shading and/or evapotranspiration. Numerous studies have demonstrated that urban parks, an essential part of urban green infrastructure, play important roles in cooling the urban thermal environment and mitigating the urban heat island effect (Huang

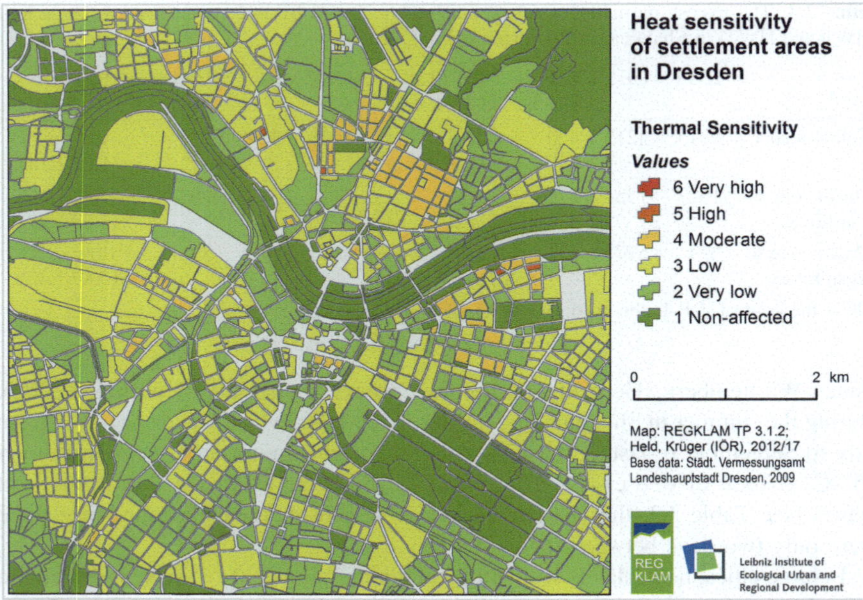

Fig. 3.4 Zones of inner-city heat islands: The map for the city center of Dresden (Germany) allows the extent to which people of all age groups are affected by urban heat islands to be read off (modified after Krüger et al. 2014)

Table 3.4 Climatic effects of urban vegetation on selected climatic elements (diverse authors in Mathey et al. 2011 and TEEB DE 2016)

Climatic elements	Climatic effects caused by growing proportion of urban vegetation	Qualitative description
Air temperature	Sinking	Increase of the shading effect, in particular by the medium and high vegetation; transpiration of the plants; reflection of solar energy; cooling below the vegetation as well as of the adjacent areas
Air humidity	Rising	Lower runoff of rainwater; increase in rainwater infiltration; increase in evaporating surfaces; increasing release of water vapor
Wind	Sinking	Emergence of temperature differences between vegetation areas and the surroundings; vertical air movements; formation of small air circulations that support or obstruct the horizontal exchange of air, according to the arrangement of the vegetation in the structure; increase in air resistance and thus decrease in wind speeds
Radiation	Unaffected to sinking	Damping of the brightness maximums under trees and higher shrubs (medium and high vegetation layer); reflection of sunlight; shading effect

et al. 1987; Taha et al. 1991; Chang et al. 2007; Bowler et al. 2010a). Many studies have shown that urban parks are cooler than their surroundings and therefore have a cooling effect like an "oasis effect" (Taha et al. 1991) or "cool island effect" (Jauregui 1990; Chang et al. 2007).

The cooling effect of urban green spaces is largely determined by the types, the area/size and the structure of the green space itself, as well as the landscape patterns of surrounding areas (Li et al. 2012; Cheng et al. 2015). The cooling effect of urban green spaces, for instance of an urban park, can be measured by the park's cool island effect (PCI), proposed by Chang et al. (2007), which is defined as the air temperature difference between the park interior and the nearby surroundings. Therefore, the park's cooling effect strongly depends on both the temperature of the park and that of its surroundings. Average cooling effects of urban parks were estimated at 0.94 °C in the day and 1.15 °C in the night by Bowler et al. (2010b) with large variations ranging from 1 to 7 °C (Chang et al. 2007; Jusuf et al. 2007). How far the park's cooling effect can reach and how large the area is that a park's cooling effect can cover are crucial issues in planning urban green spaces, as they have an influence on public health. Modeling results revealed that the cooling distance could range from 100 m (Shashua-Bar and Hoffman 2000) to nearly 500 m (Honjo and Takakura 1990; Barradas 1991) away from the edge of the park.

The higher the proportion of vegetation-covered areas, the higher the green volume and consequently also the cooling effect (Mathey et al. 2011, 2015). The modeling results for Dresden (Germany) show that the maximal temperature in summer can be lowered by large volumes of vegetation. The potentials of several urban green space types to regulate the local temperature are indicated in Fig. 3.5. Designated green spaces vary considerably in their impact on air temperature. The average cooling effect of green sites of size 1 ha ranges between 0.1 K and 2.4 K over the course of the day. Besides explicit urban green spaces, the vegetation inside settlement areas also contributes to the provision of climatic regulation effects. Thus, the highest cooling effects in settlement areas are observed when the proportion of soil sealing is low and the vegetation structure is heterogeneous, featuring grassland, bushes, as well as small and high trees (Fig. 3.6, type 3) (Lehmann et al. 2014).

The regulative effects of various types of urban green spaces vary over the course of the day (Fig. 3.6). In the daytime the climatic effects are determined by the interaction of direct solar radiation, shade, wind speed, and wind direction. Around midday it is generally possible to achieve cooling effects. However, depending on the particular structure, warmer air temperatures are encountered in the early morning and evening hours than on an asphalted reference area. Densely built-up and sealed areas are heat stores which emit heat to their surroundings at night (Lehmann et al. 2014). Due to less heating in the daytime, lower heat storage and higher evaporation, the cooling effects of urban green spaces especially in evening and night hours are much higher than those of built-up settlement areas. In the daytime, the lowering of temperatures in urban green spaces is especially perceptible in areas covered with trees. While open unsealed sites such as lawns and meadows (Fig. 3.6, type 2) generally have a high cooling potential during the night,

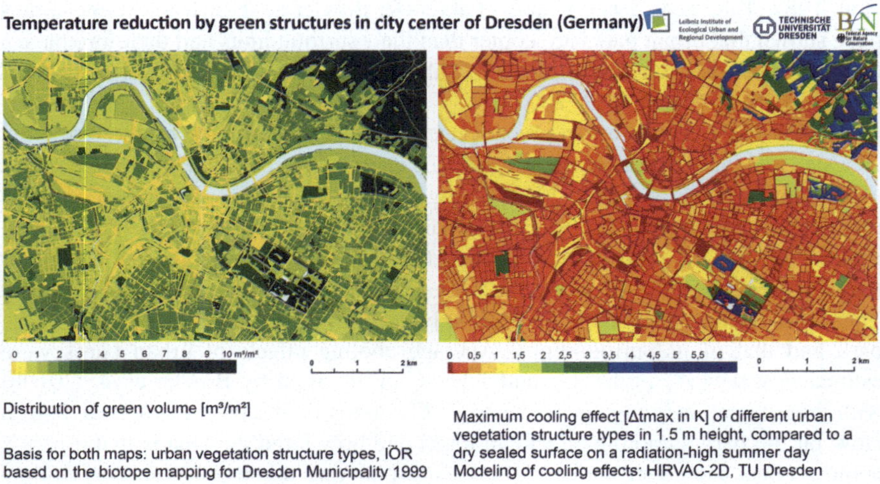

Fig. 3.5 Distribution of green volume (*left*) and maximal cooling effects (*right*) on a summer day with high radiation in Dresden (after Mathey et al. 2011, modified)

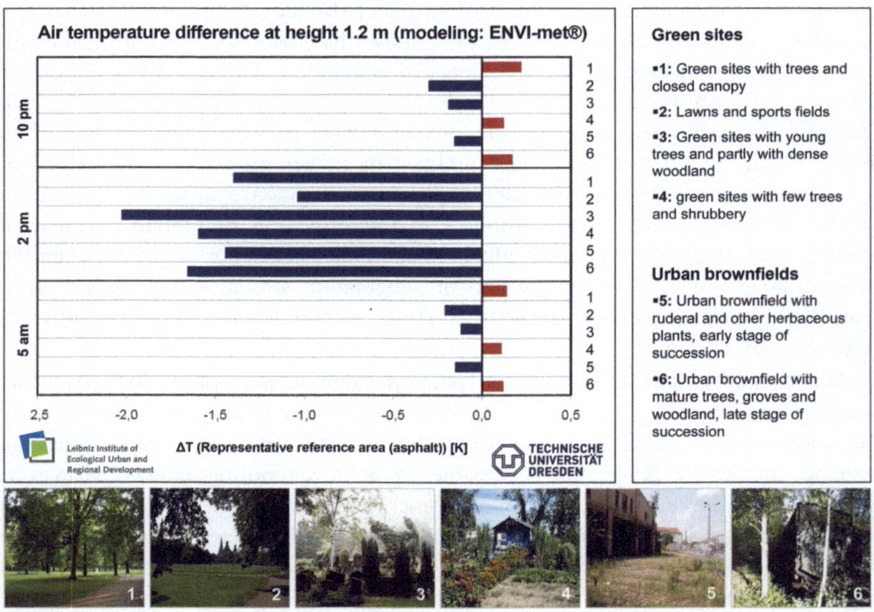

Fig. 3.6 Microclimatic effects of selected urban green space types in Dresden (Germany) at 5 am, 2 pm, and 10 pm (modeled by ENVI-met®, Bruse and Fleer 1998, modified after Mathey et al. 2015). © S. Rößler (Photo 1); R. Bendner (Photo 2, 4 and 6); S. Stutzriemer (Photo 3); B. Kochan (Photo 5)

dense woodland (Fig. 3.6, type 1) can prevent surface cooling after sundown (Mathey et al. 2011, 2015). Densely built-up, sealed quarters store the heat of hot summer days and deliver it to their surroundings in the evening hours. This overheating can be lowered by greened walls (Fig. 4.8), which have climate regulation effects. The shade of trees in parks is a popular place on hot summer days. The climatic effects of dry lawns or defoliated trees, however, are similar to the effects of sealed areas.

Besides vegetation coverage, the size of a single urban green space is of high importance as regards its potential of microclimate lowering (Table 3.5). Normally, large urban green spaces provide higher cooling effects than small ones. The doubling of the area causes a temperature reduction of 1 K; higher enlargement of the area can reduce the temperature between 1.5 K and 3 K. Perceptible climatic effects are described for urban green spaces from a size of 1 ha (Stülpnagel in Gill et al. 2007). The higher the proportion of vegetation-covered areas in the entire city, the more convenient the climatic effects on the urban climate will generally be. For example, an increase of the green space proportion by around 10% would nearly compensate the temperature rise forecast up to 2080 in the city center of Manchester. A loss of urban green spaces, from building up all urban wastelands (brownfields) for instance, would lead to a drastic rise in temperature (Pauleit 2010).

Normally, the larger an individual green space, the longer the distances that can be overcome by the cooling effects. Studies which measured air temperature based on a single large park reported that the maximum cooling distance could reach 1100 m in Göteborg, Sweden (Upmanis et al. 1998) and 2000 m in New Mexico City (Jauregui 1990), respectively. However, the climatic sphere of action of most of the urban green spaces ends, without influence of the topography, usually at distances of about 200–400 m (Stülpnagel 1987). In Shanghai, the parks' maximum cooling distance showed an increasing trend with park size with a power relationship ($R^2 = 0.734$, $p < 0.001$) between park size and the maximum cooling distance of the park (Cheng et al. 2015).

The climatic effects of urban green spaces are influenced by urban land cover and land-use change (Chen et al. 2006) and urban landscape patterns (Li et al. 2011;

Table 3.5 Size of urban parks and the occurrence of their park cool island (PCI) effect (horizontal temperature difference between air temperature of urban surroundings and air temperature in a park [K]) (compilation of diverse measurement results in Bongardt (2006) and Li's unpublished data for Shanghai)

Size class of urban park (Germany)	PCI	Size class of urban park (Shanghai)	PCI (at night)
Urban park up to 5 ha	2.9–4 K	Urban park size 1–5 ha	0.7–2.7 K
Urban park up to 20 ha	Up to 2.5 K	Urban park size 5–10 ha	0.2–3.1 K
Urban park up to 100 ha	2–2.5 K	Urban park size 10–30 ha	0.2–2.0 K
Urban park over 100 ha	1.7–6 K	Urban park size 30–150 ha	0.2–1.0 K

Mathey et al. 2011; Zhou et al. 2013). Recent studies demonstrated that landscape patterns of urban green spaces themselves influenced park air temperature or land surface temperature (LST). For example, a case study in Beijing, China showed that the percent cover of green space was a predictor of land surface temperature. LST decreased by approximately 0.86 °C with an increase of 10% green space. LST was also significantly affected by the configuration of green space, especially its patch density. The composition and configuration of green space were largely able to explain the variance of LST (Li et al. 2012; Cheng et al. 2015).

In the German Strategy for Adaptation to Climate Change the conservation of biodiversity is considered as a requirement for preserving the adaptability of natural systems. Integrative measures should use the synergies between nature conservation, climate mitigation, and climate adaptation, and at the same time they are to preserve biodiversity (Bundesregierung 2008).

3.3.2 Water Regulation, Flood Protection

Stefanie Rößler, Boping Chen, Gaodi Xie and Biao Zhang

Green spaces provide manifold regulating services regarding urban water cycles. The possibility of natural evaporation and infiltration ensures water regulation and circulation. Due to the growing challenges of dealing with the impacts of climate change, both in terms of varying precipitation schemes and of increasing heavy precipitation events, these regulating services are becoming even more relevant. Different types of green infrastructure provide different opportunities to deal with these requirements and to make use of these potentials to supplement gray infrastructure solutions.

Regulating capacities

The urban water cycle is determined by precipitation, evaporation (or evapotranspiration), infiltration, subsurface flow, surface runoff, and stream flow (Illgen 2011). These processes are influenced by a number of factors, particularly land use of the catchment area. Land cover/surface sealing has a particular influence on the direct runoff and the indirect runoff (evapotranspiration rates), which determines the retention, the infiltration capacity and the groundwater recharge (Haase 2009).

Increased building activities lead to soil sealing, which decreases the perviousness of the catchment to precipitation and the amount of vegetation. This results in a disturbed water balance and a reduction of the groundwater recharge rate. Due to the decrease in infiltration and evapotranspiration, surface runoff increases (see Table 3.6), which leads to a growing demand for technical runoff treatment and growing potentials of damage by flooding (Weller et al. 2012).

Green spaces, characterized by open soil and vegetation cover, provide a number of regulating services that mitigate the effects of increasing soil sealing by building activities. Vegetation is able to reduce the surface runoff by intercepting water

Table 3.6 Relationship between soil sealing and increase in water runoff [own chart, after Arnold and Gibbons (1996) in Paul and Meyer (2008)]

Increase of impervious surface per catchment area	Increase in water runoff (compared to forested areas)
10–20%	Twofold
35–50%	Threefold
75–100%	>Fivefold

through the leaves and stems. Before percolation, the soil can store water in pore spaces (Gómez-Baggethun et al. 2013).

By these natural processes the surface runoff is reduced, and thus the risk of urban floods is decreased and the urban sewage systems are relieved. Evapotranspiration processes are directly linked to the ability of vegetation to positively influence the microclimate. Thus, the water-related regulating services are a precondition for the provision of climate-related regulating services.

Additionally, green spaces provide areas which, in case of flooding resulting both from urban rivers as well as from the urban sewage system, offer retention/storage capacities and—in extreme situations—direct surface runoff and floods toward areas with low potential of damage compared to houses, underground car parks or underground traffic infrastructure.

Changing climate and urgent need for healthy urban water balance

A steady water balance in urban areas offers manifold benefits, from enabling natural processes regarding the soil and the vegetation growth to stabilizing urban stream systems and groundwater systems.

In view of the impacts of climate change, these benefits will become even more relevant in urban areas. Changes in precipitation regimes with respect to the amount, the duration, and the yearly distribution have several impacts in urban areas, which strongly differ between different cities and regions:

- ***Droughts*** may lead to disturbed urban water systems. The groundwater recharge might be unbalanced. Water scarcity for vegetation might lead to increasing demands for irrigation. The permeability of urban streams and thus the health of aquatic ecosystems might be disturbed.
- ***Changes in the amount and yearly distribution of precipitation*** might lead to shifting vegetation periods and require adaptation measures.
- ***Increase in heavy rainfall events*** (due to the amount and the duration of rainfall) might lead to flooding, caused by both urban streams and overloaded urban sewage systems, which means threats to human, buildings, and infrastructure.

Example 3.3 Rainwater-runoff reduction in Beijing
Based on inventory data of urban green spaces in Beijing, Zhang et al. (2012a) evaluated the economic benefits of rainwater-runoff reduction by urban green

spaces, using the rainwater-runoff coefficient method as well as economic valuation methods. The results showed that 2494 cubic meters of potential runoff was avoided per hectare of green area and a total volume of 154 million cubic meters of rainwater was stored in these urban green spaces, which almost corresponds to the annual water needs of the urban ecological landscape in Beijing. The total economic benefit was 1.34 billion RMB in 2009 (RMB: Chinese currency, US $1 = RMB6.83), which is equivalent to three-quarters of the maintenance cost of Beijing's green spaces; the value of rainwater-runoff reduction was 21.77 thousand RMB per hectare. In addition, the districts and counties were ranked according to the amount and value of rainwater-runoff reduction by the urban green areas. The results show that the average benefits per hectare of green space differed between the areas of Beijing, which may be related to the impervious surface index in the different regions (Fig. 3.7).

This study tells us that green spaces could provide significant economic value through rainwater-runoff reduction. Therefore, Beijing could profit from improved rainwater drainage through soft ground since the building and maintenance of a stormwater-drainage system would involve large costs.

Example 3.4 Ningbo Eco-Corridor—3.3 km Living Filter
The Ningbo East New Town Eco-Corridor restoration project won the 2013 ASLA Professional Award. The corridor is a wetland that runs through the east of Ningbo, a well-off Chinese city with a population of approximately 3.5 million (2010) which

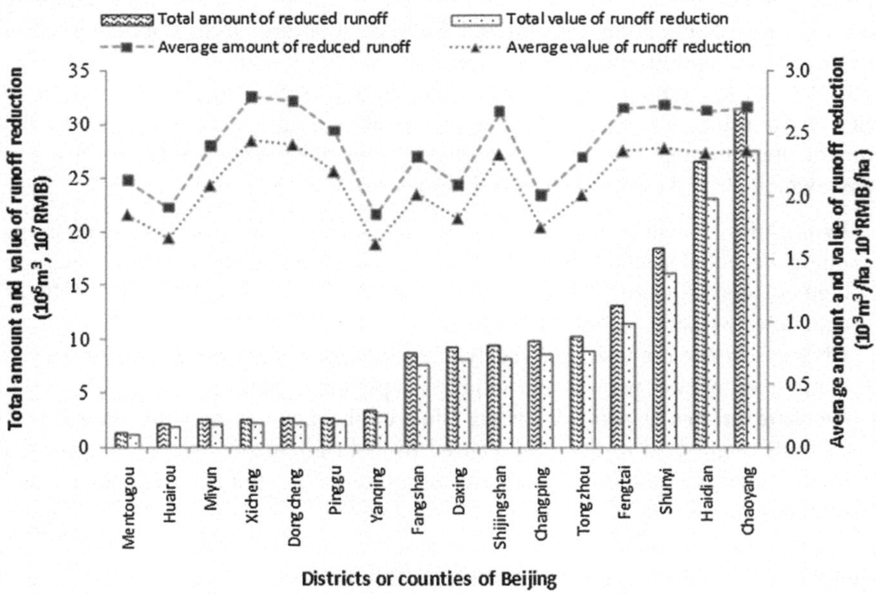

Fig. 3.7 The amount and value of rainwater-runoff reduction by green spaces in different areas of Beijing (modified after Zhang et al. 2012a)

used to suffer from significant loss of wetland and water habitats. The corridor was originally inhabitable brownfield but has now been vegetated with green spaces, and the original dead-end rivers have been reconnected with artificial canals. It now serves as a buffer zone for rainwater-runoff storage and water pollution mitigation. The eco-ponds, marsh wetlands, and aquatic habitats have also preserved the biodiversity in this area. The corridor was therefore named "eco-corridor" (GOOOOD 2014).

Many sources have been incorporated into the construction to improve water quality. The strategy is to

(1) create hills and valleys to increase water flow and initiate active water treatment;
(2) use plants and eco-methods to remove targeted existing pollutants and purify wetlands;
(3) harvest clean rainwater. According to the project developer, the water quality has already been improved from class V to class III.

Example 3.5 Qunli Stormwater Wetland Park, Ha'erbin City, China
Beginning in 2006, a new urban district, Qunli New Town, with a size of 2733 hectares, was planned to be developed on the eastern outskirts of Ha'erbin City in North China. 32 million square meters of building floor area will be constructed in 13–15 years. More than 250,000 inhabitants are expected to live here (TURENSCAPE 2011). While only about 16.4% of the available land was zoned as permeable green space, the majority of the former flat plain will be covered with impermeable concrete. The annual rainfall of this area is 567 mm, and it is concentrated in June to August (accounting for 60-70% of annual precipitation). Flood and waterlogging were frequent in this area's history.

With the restoration of an endangered urban wetland in the city center, a wetland park project was started in 2009 not only for urban greening and citizen recreation but also for mitigating the negative effects of urban floods and for urban water regeneration. The wetland is expected to perform multiple functions, including collecting, purifying, storing the rain water and maintaining ground water, improving hydrologic circulation, flood protection, and drought relief.

With the project (Fig. 3.8), the original wetland was restored and new green-blue areas such as hummocks and pools were designed around the park as buffer areas to direct, filter, and purify rain water. Moreover, rain water from the city is also directed to the wetland park through pipe systems, and before entering into the core part of the wetland it is pre-purified by the vegetation and multiple-layer green-blue infrastructure built around the wetland park. Further complementary methods include planting trees (*Betula pendula*) on the hummocks for soil maintenance and setting footpaths around the trees and wetland for citizens to enjoy the park.

It is estimated that the park will absorb the rainfall in an area of 123 hectares directly and runoff in an area of 300 hectares from the pipe network. The park's rainwater storage capacity ranges from 71,905 m^3 to 137,674 m^3 under different conditions. Moreover, there is also obviously social value in the fact that the park,

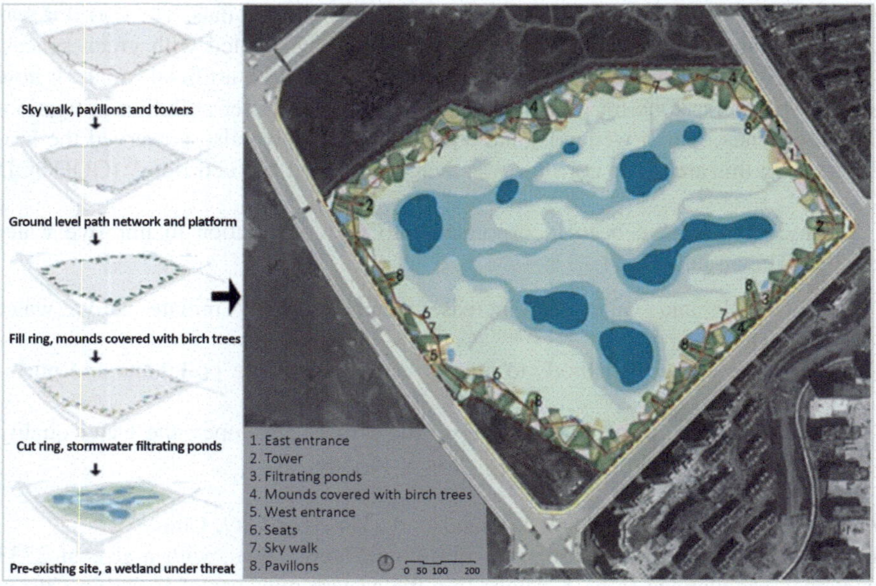

Fig. 3.8 Qunli Stormwater Wetland Park Project overview (Source: China Building Centre, TURENSCAPE 2011)

with miscellaneous forms of vegetation, serves as an ideal place for entertainment and recreation. Various outdoor activities such as picnics, photographing, and experiencing wildlife take place in the park. The park has also become a well-developed venue for leisure sports activities. For students, the wetland park also provides perfect opportunity to learn about nature and organize scientific excursions. From an economic perspective, using urban land as green "sponge body" can save large city infrastructure costs; after the project, the surrounding real estate price immediately doubled (TURENSCAPE 2011).

Example 3.6 Reducing sealed surfaces to save money
A former parking site (1.1 ha) in a residential area in Dresden, Germany, which was not being used anymore, has been redeveloped into an urban park on the initiative of the local housing cooperative as the owner of the site (Fig. 3.9). The idea was to create a kind of "urban forest" with low maintenance costs and to increase the ecological value. 5000 trees and shrubs have been planted, funded by an intervention compensation scheme of the Dresden city council. The new green space has the capacity to absorb rainwater and thus reduce the costs of wastewater removal (around 11,000 €/year). The sewage system is less frequently flooded, which helps to reduce damage and provides savings for property owners and thus also for tenants (REGKLAM-Konsortium 2013).

Fig. 3.9 The new "urban forest" in the city of Dresden. © R. Vigh/IÖR

3.3.3 Air Pollution Reduction

Biao Zhang, Gaodi Xie and Karsten Grunewald

Adverse effects of air pollution on health

Industry, residential heating, transport, and waste treatment pollute the urban air and are a major environmental concern in many major cities across the world. Polluted air is directly linked to declines in human health; in particular, it leads to increased occurrence of cardiovascular and respiratory diseases (Leiva et al. 2013; WHO 2013). The adverse effects on health due to atmospheric particulate pollution have been the subject of a growing number of studies in recent years (Tchepel and Dias 2011). For example, particles smaller than 10 μm in diameter (PM_{10}) can penetrate deep into the pulmonary passages, where any transition metals present, such as iron and copper, can release free radicals into the lung fluid and cause cellular inflammation (Birmili and Hoffmann 2006).

The health risks can vary greatly within cities. Relatively high concentrations of fine particulates exist in Western Europe in rich and structurally strong regions with high population density (TEEB DE 2016). According to Kallweit and Wintermeyer (2013), the stress caused by particulate matter results in approximately 47,000 premature deaths per year in Germany and a large number of people requiring treatment for cardiovascular and respiratory diseases. About 90% of the residents of

Fig. 3.10 December 1, 2015, typical winter smog situation in Beijing. © K. Grunewald

EU cities are exposed to air pollution levels that exceed the standards of the relevant World Health Organization (WHO) Directive (EEA 2013).

Due to the impact of rapid urban expansion, industrial development, and increased car ownership, air particulate matters have been becoming the primary pollutants in many Chinese cities, too. Figure 3.10 shows a day on which the air-quality index in Beijing and many other Chinese cities reached the top reading of 500, meaning air pollution had reached the most severe level, according to the city's environmental monitoring station ($PM_{2.5}$ reached 945 μg/m^3 and thus far exceeded the threshold value of the national standard of 75 μg/m^3). In general, the air quality has improved in China's cities and is now according to Spangenberg (2014) at levels that were usual in Germany in the 1950s. However, if city dwellers often use breathing masks it is a clear sign that the problem is far from being solved in China.

As one of the pioneering areas of reform in China, the Pearl River Delta region (PRD) has been suffering from numerous air pollution issues, including episodes of high ozone (O_3), acid deposition, and regional haze events, among others. Lu et al. (2016) applied WRF-CMAQ modeling to estimate and compare the health burden caused by four ambient pollutants (SO_2, NO_2, O_3, and PM_{10}) in the PRD during the period 2010–2013. They found that the overall possible short-term all-cause mortality due to NO_2, O_3, and PM_{10} reached 13,200 to 22,800, and the highest total economic loss was in the range from 14,770 to 25,305 million USD, which is equivalent to 1.4–2.3% of the local gross domestic product.

Mechanism of air pollutant reduction by urban green space

Vegetation removes pollutants in several ways (Rowe 2011). Plants take up gaseous pollutants through their stomata, intercept particulate matter with their leaves, and are capable of breaking down certain organic compounds such as poly-aromatic hydrocarbons in their plant tissues or in the soil. In addition, they indirectly reduce air pollutants by lowering surface temperatures through transpirational cooling and by providing shade, which in turn decreases photochemical reactions that form pollutants such as ozone in the atmosphere. For example, Akbari et al. (2001) reported that when maximum daily temperatures in Los Angeles were less than 22 °C, ozone levels were below the California standard of 90 parts per billion; at temperatures above 35 °C, practically all days were smoggy. Moreover, the cooling effect of urban green spaces in Beijing reduce the need for air conditioning, and a lower energy consumption results in lower emissions from power plants (Zhang et al. 2014).

Trees remove gaseous air pollution primarily by uptake via leaf stomata, though some gases are removed by the plant surface. Once inside the leaf, gases diffuse into intercellular spaces and may be absorbed by water films to form acids or react with the inner leaf surface. Trees also remove pollution by intercepting airborne particles. Some particles can be absorbed into the tree, though most particles that are intercepted are retained on the plant surface. The intercepted particle is often resuspended into the atmosphere, washed off by rain, or dropped to the ground with leaf and twig fall. When air with pollutants flows onto the tree crown, the wind is slowed in the forest and the particulate matter drops down owing to its gravity. The blocking of air pollutant by urban forests is very effective for TSP (Total Suspended Particles) and PM_{10}, but unclear for $PM_{2.5}$ (Chen et al. 2003; Ren et al. 2006).

Consequently, the leaves only absorb fine and superfine particulate matter (Sæbø et al. 2012; Zhao et al. 2014). Vegetation is only a temporary retention site for many atmospheric particles.

Potential for air pollutant reduction provided by urban green spaces

People have known for a long time that trees can help to reduce air pollutants. The Roman Senate recognized the value of orchards in villas surrounding the city of Rome for maintaining air quality and forbade their conversion into urban housing (Cowell 1978).

Nowak (2006) recently studied air pollution removal and air-quality improvement by urban forests for several cities in the United States. They reported that the total air pollution removal (O_3, PM_{10}, NO_2, SO_2, and CO) by urban trees amounted to 711,000 metric tons. Using assumed urban forest structure values such as the leaf area index, the mean removal of PM_{10} by trees in Los Angeles, United States was estimated at 8.0 g/m^2. German and Dutch studies estimate the filtering capacity of trees at 5–15% (Langner 2006; Kuypers et al. 2007 in TEEB DE 2016). Thus, urban green space can effectively complement other measures to improve air quality, such as reduction of industrial emissions or traffic regulation (Vos et al. 2012).

There were 2.4 million trees in the central part of Beijing in 2002, the total annual air pollutant uptake by these trees from the atmosphere was around 1.3 million kg, the standardized pollution removal rate was 27.5 g/m^2 (canopy cover), and PM_{10} accounted for 61% of total air pollutants (Yang et al. 2005). Kremer et al. (2016) estimated the air pollution removal by trees, shrubs, and grass for New York City at 2.8 million kg annually. Pollution reduction by trees is most important, but other vegetation structures can contribute to improving the air quality in cities as well. Derkzen et al. (2015) offer some examples of pollution reduction service for three selected green types (air purification in gm^{-2} a^{-1}): woodland: 2.7; shrubs: 2.1; herbaceous: 0.9. However, these numbers only provide an orientation. They vary with respect to the site conditions and cannot be easily transferred.

Green roofs and façades can also act as passive filters of airborne particulate matter. While not as effective as street trees, due to lower surface roughness lengths and increased distance from sources, they can be considered for remediation of urban air pollution because their construction does not require major upheaval of the urban built environment, as tree-planting schemes often do. There is already some evidence for the potential of green roof vegetation for air pollution removal. Yang et al. (2008) found that 1675 kg of air pollutants, such as NO_2, SO_2, and PM_{10}, were removed by 19.8 ha of green roofs in one year. A study in Toronto found that 58 metric tons of air pollutants could be removed if all the roofs in the city were converted to green roofs, with intensive green roofs having a higher impact than extensive green roofs (Currie and Bass 2008). The study of Speak et al. (2012) verified that 0.21 tons of PM_{10} a year were removed from Manchester city center in a scenario involving all flat roofs within a chosen area being installed with an extensive green roof; this is the equivalent of 2.3% of the PM_{10} inputs of this area.

Influence factors of air pollution reduction

Pollution removal values for different plants vary among tree species based on the amount of tree cover, pollution concentration, length of in-leaf season, precipitation, and other meteorological variables that affect tree transpiration and deposition velocities. All of the factors combine to affect total pollution removal and the standard pollution removal rate per unit tree cover.

The leaf characteristics will influence the deposition of air pollutants on leaf surfaces. Tree species having deep channels or dense hairs on the leaf surface (e.g., *Populus alba×berolinensis*, *Prunus maackii*) had greater effects of dust removal, while tree species with strumose projections on the leaf surface (e.g., *Pyrus ussuriensis*, *Prunus padus*) had weaker effects (Chai et al. 2002). Usually evergreen trees have higher efficiency in removing air pollutants because of their longer time of foliage retention. Growth rate influences the size of the functional crown surface for air pollutant removal, and fast growing trees can usually provide a surface for air pollutants soon after they are planted.

The velocity of wind can affect the pollutant removal capacities. For instance, Wang et al. (2006) found that wind with a speed of 10.4 m/s cannot blow away the particulate matter retained on the leaves of *Platycladus orientalis*, *Sabina Chinensis*, *Pinus tabulaeformis*, and *Picea koraiensis*.

Estimated values of the ecosystem service air pollution reduction

Values and benefits regarding the impact of urban green on air quality have been recognized at different spatial scales (examples in Box 3.2).

Box 3.2 Examples of air-quality capabilities and benefits of urban green space

Barcelona (Spain): 166 tons of particulate matter (PM_{10}) are removed by urban green space annually (22% of total city dust emissions); the monetary value amounts to annually 1.1 million USD (Baró et al. 2014).

Beijing (China): Feng et al. (2007) measured the dust absorption capability of different plant leaves in the Mentougou District of Beijing and evaluated the amount and value of dust absorption by this natural vegetation by RS and GIS means. They found that shrub forests and deciduous broadleaved forests were the main providers of dust absorption services, the natural vegetation in the Mentougou District could reduce dust by 2.95×10^5 t per year and bring a benefit of 6.71×10^7 RMB. Tong et al. (2015) have taken the Beijing urban area with serious traffic pollution as a case study and estimated the total amount of $PM_{2.5}$ removed annually by road green space, as well as its health benefits. Their results showed that the road green space in the Beijing urban area removed 1.09 tons of $PM_{2.5}$ per year and reduced the population's health risk significantly. Consequently, it obtains health-related economic benefits amounting to 198 million RMB.

Chicago (USA): Removal of O_3, PM_{10}, NO_2, SO_2, and CO_2 from the air by trees results in an annual monetary benefit of 6.4 million USD (Nowak et al. 2010).

Guanzhou (China): Wu et al. (2009) studied the functional value of absorption and removal of atmospheric pollutants by urban forests in Guanzhou. The results showed that the purification capacities of different types of urban forests ranged from 36.84 kg/ha to 365.28 kg/ha per year; the total economic value of SO_2 absorption amounted to 1.13×10^8 RMB; the urban forests were able to absorb 3078.52 tons of fluorine per year, equivalent to 3240.45 tons of hydrofluoric acid per year; the functional value was 8.10×10^4 RMB/a; the total amount of dust capture was 6.08×10^4 t/a; and the economic value was 1.91×10^4 RMB/a. Therefore, the total value of air pollutant removal by urban forests in Guangzhou city was 1.14×10^9 RMB/a.

Because plant species possess varying abilities to remove air pollutants and reduce emissions, they can be selected to maximize improvements in air quality. For example, evergreen conifers may provide a greater benefit than deciduous trees with respect to particulate matter, ozone, NO_x, and SO_x, which occur mainly while plants are actively growing and in-leaf. Individual plant species also exhibit vast differences in their ability to take up pollutants. Part of the urban tree canopy is heterogeneous with small patches or individual trees, and this mixed canopy effect would tend to increase pollutant deposition. The US Environmental Protection Agency has introduced urban tree cover as a potential emerging measure to help meet air-quality standards (US EPA 2004). In short, policy options that take into consideration the full functionality of urban forests as well as their spatiotemporal heterogeneity, scale, proper species selection, maintenance, water use, volatile organic compound (VOC) emission rates, allergenic effects, spatial arrangement, and quantity and quality will maximize the ambient air quality and human well-being of the urban populace.

3.4 Cultural Services of Urban Green Space with a Focus on Recreation

Karsten Grunewald and Jürgen Breuste

The Millennium Ecosystem Assessment (MEA 2005) defines Cultural Ecosystem Services (CES) as "nonmaterial benefits people obtain from ecosystems through spiritual enrichment, cognitive development, reflection, recreation, and aesthetic experience." Within the framework of ES, all values have to be assessed from a human perspective. This is not an easy task, because the demand of people is strongly dependent on cultural conditions. Personal aims as well as public opinion, which is strongly influenced by the media, are important factors in determining these values. Assessing CES has been considered one of the most difficult tasks in ES research. This results in particular from the fact that unlike other categories of services, they are of an intangible nature (Grunewald and Bastian 2015).

Nevertheless, there are indications and interrelations with physical characteristics of urban landscapes. Urban nature provides manifold opportunities for enjoyment, inspiration, intellectual enrichment, aesthetic delight, and recreation. Such "psychosocial" services are no less important for people than are regulation and provisioning services; however, they are often neglected or not fully appreciated. One reason is the difficulty of valuating them economically, especially in monetary terms. A second group includes "information" services, i.e., the contribution of ecosystems to knowledge and education. In order to gain information on CES, qualitative (see Box 3.3) and quantitative valuations (see below) can be applied.

Box 3.3 Perception of CES provided by urban green in Berlin
Maraja et al. (2016) investigated perceptions and understandings of CES and their benefits in the urban context of Berlin. They found a wide variety of perceptions on CES, and that the categories often overlap. In the interviews (n = 2536) by the authors, the most important CES groups in Berlin were (frequencies of aggregated CES in percentage):

- Values for recreation and tourism (30.0%)
- Values for nature awareness (12.2%)
- Aesthetic values (12.1%)
- Values for sense of place and regional identity (9.9%)
- Values for social relations (8.5%)
- Educational values (6.6%)
- Spiritual and religious values (6.4%)
- Values for cultural diversity (5.3%)
- Cultural heritage values (5.2%)
- Values for social and motoric development (2.5%)
- Values for inspiration (1.4%)

Description of the CES "recreation in the city"

Highly compact cities are a stressful environment for their citizens. In general, the speed and number of impressions cause hectic lifestyles with little room for rest and contemplation (Bolund and Hunhammer 1999). Against this background, recreation opportunities may be the greatest perceived benefit of urban green space (Andersson et al. 2015). As pointed out in Sect. 3.1, spending time in green areas results in improved physical and mental health, and cities become more attractive when they offer ample green space with public outdoor recreation opportunities.

Ecosystems in cities deliver services for the residents regarding the experience of nature, recreational activities, and aesthetics (Barbosa 2007; Kabisch and Haase 2013; Elmqvist et al. 2015; Lee et al. 2015). In the following, we will refer to these services as the CES "recreation in the city." In considering the provision of the population with recreational areas, their reachability (pedestrian distance) and public accessibility are important factors, in addition to their percentage of the area (referred to the entire city area or to the number of inhabitants) and quality (Comber et al. 2008; Kabisch and Haase 2014; Wolch et al. 2014).

In the EU's CICES classification (Common International Classification of Ecosystem Services, Haines-Young, and Potschin 2013), the CES "recreation in the city" belongs mainly to the classes "Experiential use of animals, plants, and land-/seascapes in different environmental settings" and "Physical use of land-/seascapes in different environmental settings" in the group "Physical and experiential interactions." It is part of the division "Physical and intellectual interactions with biota, ecosystems, and land-/seascapes in different environmental settings" and the section of cultural services. However, other classes of cultural ES are concerned as well (e.g., aesthetic). In a narrower sense, it mainly comprises "daily or leisure-time recreation" and "recreation in the residential environment."

The proximity of green spaces is a key criterion for choosing a place of residence (TEEB DE 2016). Recreational opportunities in public settlement areas are key to improving the quality of life for people without a motor vehicle of their own and especially for population groups of lesser mobility such as the elderly, the disabled, and children. The accessibility of recreational spaces for all population groups is in particular also relevant for social justice (Panduro and Veie 2013; Kabisch and Haase 2014; Wüstemann et al. 2016).

We therefore postulate as a goal for the CES "recreation in the city" that publicly accessible green spaces should be reachable in the neighborhood for every resident (Grunewald et al. 2017a). Availability of urban green space can be understood as the amount of urban green area at a certain defined distance from where urban residents live (Kabisch et al. 2016).

Important measurable parameters that describe the ES "recreation in the city" are:

- Share of green space in the settlement area with the potential to provide recreation (supply in the sense of "provision by nature"), degree of provision and quality of the green space;
- Settlement areas with residential use, inhabitant data, distances between residential areas, and proximate green spaces (for estimation of demand aspects).

Evaluation of the recreation services of urban parks and forests

Example 3.7 Shanghai parks

The space of urban public parks has nearly doubled to 1521 ha in 2005 over the course of only 15 years (Sect. 4.3.3). Urban parks appear to play a vital role for Shanghai's residents with respect to quality of life. On average, each district spends 26 RMB per m^2 and year for the maintenance of parks. Shanghai parks are highly frequented and very much valued by its visitors (Dong 2006; Zippel 2016).

In traditional Chinese thinking, the main purpose of green spaces was to fulfill the demand for passive recreation. Much emphasis was placed on the arrangement, composition, structure, and color of plants. Parks were meant to serve as places for relaxation, social interaction, cultural events, and sports activities (Jim and Chen 2006).

Several studies show that most of the visitors come from the neighborhood. For a great majority of the visitors of three selected parks in Shanghai (Kai Qiao, Tianshan, and Zhongshan Park), the environmental conditions are an important reason for visiting the park (83–94%). The majority of visitors questioned also expressed the view that nature experience is an important reason for visiting the park (73–86%). The ornamental nature of the parks creates a general vision of nature for most of the urban dwellers (Breuste et al. 2013a, b).

Questioning (n = 106 persons) and observations of user frequency by Zippel (2016) in 2014 in Fuxing Park, Shanghai's oldest park, showed that the majority of interviewed visitors (64.4%) comes from the neighborhood (less than 30 min walking distance). More than half of the participants (53.3%) liked to spend time in the park during the morning, followed by people who preferred the afternoon (21%). A total of 89.3% valued the park as either important or very important. The majority (58.1%) of the respondents chose relaxation as their principal visiting purpose (Table 3.7). The survey results indicate that visitors are generally satisfied with the park structures provided, have a close relationship with the park and its

Table 3.7 Motives for park visits of survey respondents in Fuxing Park (Zippel 2016)

Motive for park visits	N (321)*	% of respondents	% of responses	Motive for park visits	N (321)*	% of respondents	% of responses
To enjoy fresh air	65	20.2	69.9	To use facilities for children	12	3.7	12.9
To be in nature	58	18.1	62.4	To listen to music	11	3.4	11.8
To exercise	37	11.5	39.8	To get new energy	10	3.1	10.8
To enjoy tranquility	32	10.0	34.4	To have social interaction	9	2.8	9.7
To learn from nature	20	6.2	21.5	To observe other people	6	1.9	6.5
To enjoy the scenery	19	5.9	20.4	To dance	4	1.2	4.3
To have private space	19	5.9	20.4	To make music	1	0.3	1.1
To distract from problems	18	5.6	19.4	Others	–	–	–

*Respondents had the option to mark maximally four out of 16 given responses. In total, 93 participants filled out the question correctly, with the result that 321 responses could be counted

services and mostly need the park as the only option in their neighborhood to enjoy fresh air, to relax and to be in nature.

The findings obtained provide guidance and proposals for Chinese designers and municipal authorities to efficiently plan and manage parks with people's interests in mind and to improve the cultural ES (Zippel 2016).

Example 3.8 Beijing's forest resources
Beijing's forest resources have been increasing significantly since the 1950s as a result of active planting and management. In 2007, the forest area of the municipality reached 1.1 million hectares (Wu et al. 2010). The dominant tree species include *Quercus mongolica*, *Platycladus orientalis*, *Pinus tabuliformis*, *Populus davidiana*, *Betula platyphylla*, *Robinia pseudoacacia*, and *Larix principis-rupprechtii*. The forests are rich in biodiversity, hosting a variety of fauna and flora. Wu et al. (2010) estimated the value of forest ecotourism using the travel cost method for 11 forest parks in Beijing. The total flow value of annual output of forest goods and services of Beijing city was 47.9 billion yuan, of which 2.2 % were accounted for by sociocultural benefits (1.04 billion yuan).

Example 3.9 Value of urban woodland in Munich
Lupp et al. (2016) assessed the actual on-site recreational use of two forests in the north of the Munich metropolitan region using camera traps for visitor counting. Interviews were used to gain additional data for both the recreation quality and the importance of various CES for forest visitors, but also to better understand the

catchment areas of the two forests and to calculate a monetary value (using the travel cost method and a “day ticket” approach).

Visitor numbers in the investigated forests were much greater than rough estimates had suggested; visitors also had quite long traveling distances to the forests. The average distance between home and the forest was 29.7 km. Jogging or Nordic walking were shown to be important recreational activities. In some of the monitored locations, almost half of the recreationists carried out these sports. Depending on the method chosen, the calculated monetary value of recreation reached up to 15,440 Euro per hectare per year.

Implementation of the national indicator “accessibility of urban green space” in Germany

In Germany, a national indicator “accessibility of urban green space” was developed in the context of the EU Biodiversity Strategy 2020 (Target 2, Action 5), and it was recommended to integrate the indicator into the German National Strategy on Biological Diversity. In this context, the potential of urban green space for everyday outdoor recreation of short duration, such as walking, physical exercise, and relaxation, was considered. For methodological details of indicator development and calculation see Grunewald et al. (2017a).

The result shows that nationwide in 2013, 74.3% of the inhabitants of the German cities (all cities with more than 50,000 inhabitants) were able to reach at least green spaces (>1 ha) at a linear distance of no more than 300 m (≈500 m walking distance) and larger green spaces (>10 ha) at a maximum linear distance of 700 m (≈1000 m walking distance). This means that 25.6 million city dwellers have access to both types of green spaces in German cities. However, around 6.1 million inhabitants do not have one of the two types of green spaces in their residential environment yet.

For example, the city of Stuttgart has an outstandingly high percentage of accessibility of green spaces (80%) for the residential population, even though the area of green spaces per inhabitant is relatively low (116 m^2/inh.). The public parks in the densely populated core city and the remaining green spaces on the adjoining slopes as well as high population densities in the centers surrounded by forest areas outside the city center are among the special features in Stuttgart (Grunewald et al. 2017b).

The indicator is a rather simple, robust, and reproducible measure. It exhibits an average value for Germany, but also allows for simple nationwide comparisons between the cities. The proposed indicator is easy to interpret, since the closer the degree of provision comes to 100%, the higher the welfare effect (Krekel et al. 2015). This target value is easier to justify, to compare and to communicate than the green space provision (area per inhabitant). The number of inhabitants is more useful as a reference quantity for examining the accessibility of green spaces than the municipal area, since a concentration of population has a stronger influence on the indicator, which therefore exhibits a closer relationship to the people looking for rest and recreation. Comparisons with other national and international assessments show that the results are plausible. However, it only makes sense to compare numbers if they have been calculated with similar methodology and comparable databases.

Conclusion

According to a review conducted by Hernández-Morcillo et al. (2013), approximately 70 cultural service indicators have been developed so far. The majority of them, however, only refer to selected benefits generated by ecosystems, and particularly to recreation and ecotourism. The majority of CES subcategories are imperceptible in nature. Moreover, aesthetic and spiritual services are difficult to express in quantitative terms, as they differ across cultures and individuals. Therefore, comprehensive CES assessment is based on descriptive information rather than quantitative data, or on combined data as shown above.

It is important to focus not only on green space quantity but also on accessibility and quality (e.g., recreation facilities, beauty and harmony, safety etc.). Providing high accessibility of public urban green space is already a principal goal in planning and also in compact cities (Sect. 5.2). But as our assessments have shown, it is far from being reached in German or Chinese cities, although green space provision of a city in total or per inhabitant might be high.

In Chinese megacities, urban parks are becoming very important for CES supply, but the Western ideal of the park is relatively new in China (Shi 1998). The urban parks created in Beijing and Shanghai during the early twentieth century were reserved for Europeans, wealthy merchants and dignitaries, while commoners were actively excluded (Wolch et al. 2014). This changed in the last decades. Opening hours of parks were extended and access improved. Access to green space is also an environmental justice issue in Germany and in particular in China due to high residential densities, historical patterns of urban development, and little private green (e.g., front gardens).

3.5 Provisioning Services

Martina Artmann

Provisioning ES involve resource outputs supplied by the ecosystem, such as fresh water (e.g., for drinking water and energy), raw materials (e.g., timber, wood, cotton), or food (e.g., from plants, animals) (MEA 2005; TEEB 2011). According to a review, water and food supply are the most important provisioning ES investigated in current research focusing on cities (Haase et al. 2014). These ES can be supplied within the city for instance through urban agriculture, urban gardens, urban forests, lakes, and streams. However, in particular for provisioning ES the teleconnections of cities with their hinterland and more distant places become obvious. Most cities need to receive water from distant places to fulfill their demand, and therefore cities can be regarded as receiving systems of water but also of food (Yang et al. 2016).

Relevance of Provisioning ES for Cities in China and Germany

Today the satisfaction of human needs for water availability is a crucial challenge. Research shows that currently 150 million urban residents are suffering from perennial water shortage. Due to ongoing worldwide urbanization processes and climate change, this number is expected to increase to nearly 1 billion people (McDonald et al. 2011). Although in Germany the water supply is in general secured, climate change has increasing impacts on it as well. The increasing frequency of dry summers leads to dehydration within the urban water cycle, threatening the local water supply—also in qualitative terms. Due to the heating of surfaces, drinking water pipes and the distribution network warm up. The high temperature increases the risk of recontamination of the drinking water, reducing the water quality. Urban green infrastructure and its proper management can help to secure the fresh water supply. German environmental agencies suggest to shade the surfaces with trees so as to reduce the heating of the pipes. Moreover, urban forests can improve the quality of the drinking water. The potential of different tree species to contribute to water purification capacity can be estimated using the ecosystem service approach (TEEB 2011).

Increased access to safe drinking water is of especially high importance in China. In Beijing, the per capita availability of renewable water resources is one of the lowest of the world. Moreover, the natural water inflow to Beijing decreased by 91% between 1956 and 2014 (Yang et al. 2016). Through technical solutions, lacking ES can be transferred to cities. For instance, China's ongoing South–North Water Transfer Project aims to transfer water from the Yangtze River basin from southern to northern China, in particular for urban–industrial and residential purposes (Dong and Wang 2011). However, teleconnections of water dynamics have negative environmental (e.g., degradation of riparian and aquatic ecosystems) and socioeconomic effects (e.g., costs for restoration of ecosystems, resettlement of people) (Yang et al. 2016). These examples show the vulnerability of key ES such as water supply due to human impacts on the ecosystem. Therefore, economic development must go hand in hand with environmental protection to secure human well-being and sustainable urban development.

To approach sustainable urban development, cities should steer toward an increase in the self-sufficiency with respect to food as well, which has social, ecological, and economic benefits. Urban food supply such as through urban agriculture and urban gardening can support access to healthy food, decreases environmental impacts through local production, and fosters the local economy and social cohesion (Grewal and Grewal 2012). However, valuable soils and agricultural areas are lost due to urban sprawl, as shown on the example of Leipzig (Germany) (Artmann 2013) and Nanjing City (China) (Zhang et al. 2007).

Through conserving, developing, and (re-)vitalizing urban green spaces by planting edible plants, cities can contribute to local urban food supply. It is estimated that cities located in moderate climates can supply 30% of their needs for fruits and vegetables (Whitfield 2009). Urban food is mainly produced in managed agro-ecosystems. However, also forests, rooftop gardens, community gardens as well as marine and freshwater systems contribute to the food supply for humans (Grewal and Grewal 2012; TEEB 2011). Due to limited space in cities, there is a need to reclaim further spaces for food production, such as roofs or public spaces. In Germany, for instance, it is estimated that 300,000 m^2 of roof areas are unused although they could be developed for food production (Rauterberg 2013). The concept of the “edible city” aiming to use public green spaces for local food production is gaining more and more attention from urban planners worldwide. The first German city which implemented the concept into practice was the city of Andernach. The example of Andernach showed that the increase in local food supply from public spaces has synergies with other ES and socioeconomic co-benefits, such as an increase in recreational values of green spaces, an improvement in the use of regional cycles of materials, support for participative urban planning, and a reduction of costs for green space maintenance. To secure food supply in China, there is especially the need to limit the loss and degradation of cultivated land due to urbanization and industrialization. It is argued that the “(…) accelerated urbanization and industrialization will decrease cultivated area to the point where China can no longer feed its people in the next decades” (Chen 2007). The assessment of the value of ES can show up crucial lacks of resources according to humans’ demands due to degrading ecosystems’ potentials to sustain ES.

Assessment of provisioning urban ecosystem services

Compared to cultural and regulating ES, provisioning ES are easier to quantify, especially in monetary terms. Thus, food or raw materials already have a market price. A monetary valuation of provision services is carried out by making market observations and screening reference prices (TEEB 2011). A case study in China did a monetary assessment of material production and water supply, among other ES, in the Sanyang wetland, which is a degraded river wetland near Wenzhou city. The water supply was quantified by the water price provided by the local waterworks, with its supply being considered to depend on the water quality. Material supply was assessed mainly based on agricultural production by calculating the market price minus the cost for production per product unit. The results showed that the potential value of the Sanyang wetland was 55,332 yuan ha^{-1} $year^{-1}$. However, the calculation of the current value only yielded 5807 yuan ha^{-1} $year^{-1}$. Thus, there is a need to restore 89.5% of the ES to tap the full potential of the wetland’s value, which is disturbed for instance by water nutrient enrichment and accumulation of heavy metals (Tong et al. 2007).

Besides the monetary valuation, provisioning ES are also assessed in nonmonetary terms. For instance, Kroll et al. (2012) evaluated the ratio between the supply and demand of water and food provisioning in the region of Leipzig-Halle (Germany) between 1990 and 2007. The assessment of the ES is based on land-cover patterns, with different indicators being applied: The water supply was calculated using groundwater recharge, the water demand due to the groundwater consumption of the vegetation and water consumption of different sectors (e.g., industry, households, agriculture); the food supply was quantified by food production per unit of different land-use types (e.g., arable land, forests, water); and the food demand was measured by the average food consumption per capita in Germany at the municipal level. The results showed serious impacts by urbanization. Due to an increase of urban, industrial, commercial and transportation areas, the water supply decreased. The impacts could be mitigated through decreasing water demand in the urban, industrial, and commercial areas. Urbanization also threatens the food supply. Due to expansions of urban areas and an increase in biofuel crop cultivation, fertile soils and arable land got lost, reducing the food production. However, it is argued that through increased productivity and increased land-use intensity the loss of arable land can be counteracted to secure a sufficient supply of food according to the residents' demand (Kroll et al. 2012).

3.6 The Economic Benefits of Urban Biodiversity and Ecosystem Services

Henry Wüstemann and Dennis Kalisch

The previous sections have highlighted the high variety of benefits provided by urban biodiversity and ecosystems. This particular section addresses the economic value of these multiple benefits by presenting valuation studies from Germany and China. Over the last decade a number of case studies from China and Germany have been published, all underlining the economic value of urban biodiversity and ecosystem services (ES). Besides several limitations, the economic valuation of changes in ES provision and biodiversity affected by particular policies has two advantages: First—depending on the applied valuation method—it allows the benefits of a bundle of ES to be assessed. Second, monetary valuation allows these benefits to be compared with the costs for future policy interventions, since both benefits and costs are expressed in the same units. Moreover, demonstrating the economic value of ES enables policymakers to consider the total costs of using natural resources and supports the decision process for different planning alternatives.

The underlying concept of the economic value of biodiversity and ES is the Total Economic Value (TEV), which comprises use and nonuse values. Various economic methods are suitable for covering these use and nonuse values; they can be roughly divided into two groups: *Revealed Preference Methods* (RPM) and *Stated Preference Methods* (SPM). RPM such as the *Hedonic Price Method* (HPM) or the *Travel Cost Method* (TCM) try to infer the value of a nonmarket good by observing the actual behavior of individuals on related markets (Alriksson and Öberg 2008). In the case of house prices, the overall assumption of the HPM is that these prices are affected by various variables (Melichar et al. 2009). According to Bateman (1993), the HPM depends on several assumptions: First, individuals are able to perceive environmental quality changes and they are willing to pay for these environmental quality changes. Second, the entire study area is treated as one competitive market with perfect information regarding house prices and environmental characteristics. Third, the housing market is in equilibrium. The HPM is also subject to some limitations relating to problems of information asymmetries, individual perception, subjectivity, continuity, aversion behavior, market segmentation, and the assumption of equilibrium (Vanslembrouck et al. 2005). However, if the assumptions of the HPM mentioned above are properly taken into account, the method can reveal valuable information about the value of environmental attributes (Bateman 1993).

In the group of SPM, the respondents are directly asked about their preferences for a hypothetical transformation of the environmental goods considered (Bateman et al. 2002). Both the Contingent Valuation Method (CVM) and the Discrete Choice Experiments (DCE) belong to the SPM. The standard CVM has been widely used in landscape valuation since many years (e.g., Santos 1998).

An alternative way to monetize environmental amenities is cost-based valuation. Cost-based approaches are based on the calculation of costs associated with the provision of ES through artificial means (Garrod and Willis 1999). Cost-based approaches include avoided cost methods, replacement cost methods and mitigation, and restoration cost methods (TEEB 2010). In general, cost-based methods are applied to the valuation of regulating services such as climate mitigation benefits and water and air purification benefits. However, limitations of cost-based approaches mainly arise from the absence of markets for ES and limited understanding of the cause–effect linkages between the ES being valued and the marketed commodity (Spash 2000).

Another method often applied for the economic valuation of urban green space (life satisfaction/happiness) is the *Marginal Rate of Substitution* (MRS). Based on the assumption that life satisfaction data are an approximation of what Kahnemann and Sudgen (2005) labeled "experienced utility," this estimated relationship is used to derive the implicit MRS between income and the environmental amenity in question (Bertram and Rehdanz 2015). Applications of happiness data for the valuation of environmental amenities include among others urban green space (Krekel et al. 2016), noise (Van Praag and Baarsma 2005), and air pollution (Welsch 2006).

Available studies valuing the economic benefits of urban ecosystems and biodiversity in China and Germany treat the valuation of urban green using HPM (e.g., Kolbe and Wüstemann 2014) and Life Satisfaction Approaches (Krekel et al. 2016; Bertram and Rehdanz 2015), Contingent Valuation (Jim and Chen 2006), Choice Experiments (Bertram et al. 2017), and climate regulation and mitigation with cost-based approaches (Aevermann and Schmude 2015). In the following section, results from ES valuation studies from both China and Germany are presented in detail to give a brief overview of the direction of contemporary research.

Climate regulation benefits: Urban parks regulate microclimate within cities by contributing to a reduction of the "urban heat island effect" (Sect. 3.3.1). Higher surface temperatures in urban areas influence the mortality rate of urban dwellers. Tan et al. (2010) demonstrate that summer mortality rates in and around Shanghai are related to the level of urbanization. Heat-related mortality is significantly higher in the city center compared to suburban areas (e.g., during the 1998 heat wave, the excess mortality rate in the urban area was about 27.3/100,000 compared to only

7/100,000 in the suburban districts). The regulation service of urban green can also be valued by considering cost savings for cooling energy. To give an example, Leng et al. (2004) analyze the nature-based heat stress reduction potential and microclimatic regulation function of urban forests in Beijing. The annual value is RMB 93.5 $\times$ 10^6 (1 € = RMB 10.8 in 2008), under the assumption that energy for cooling is needed on 100 days per year.

Climate mitigation benefits of urban forests: Urban vegetation and its underlying ES contribute to the balance of greenhouse gases by carbon sequestration and storage. Studies estimating the climate mitigation potential of urban trees in Germany are rare, as the majority of studies focus on carbon stocks of national forests (e.g., Dieter and Elsasser 2002). Aevermann and Schmude (2015) quantify the economic value of the ES "carbon sequestration and storage" for an urban park in Munich. The economic value of the ES per year is around 15,000 € for carbon sequestration and more than 670,000 € for carbon storage. In China, urban forest valuation is a new but rapidly growing branch of environmental research (e.g., Jim and Chen 2009). Green infrastructure in Chinese cities plays a crucial role in climate change mitigation, as the total amount of carbon stored in the vegetation of the urban green infrastructure of 35 cities is estimated at 18.7 million tons, with an average carbon density of 21.34 t/ha (Chen 2015). Approaches for the economic valuation of climate protection benefits of ecosystems and biodiversity include among others abatement and damage costs as well as market prices from carbon emission trading systems (Hartje et al. 2015).

Life satisfaction and residential well-being: Urban green space contributes to residential well-being and health (Sect. 3.1). The indicator "Satisfaction with Life" is usually obtained from an eleven-point single-item Likert scale which asks "How satisfied are you with your life, all things considered?" (Krekel et al. 2016). The Marginal Rate of Substitution (MRS) is an approach that can be used for the economic valuation of urban green space based on life satisfaction data. In the German context two studies (Bertram and Rehdanz 2015; Krekel et al. 2016) exist which apply the MRS for the economic valuation of urban green space. The study by Bertram and Rehdanz (2015) estimates an implicit MRS of 26.82 Euro per person per hectare per month based on average green space availability and average income. The only nationwide study which applies the MRS for the economic valuation of green space is the study by Krekel et al. (2016). The study finds that, on average, residents are willing to pay 23 € of monthly net individual income in order to increase the amount of green space in a 1 km buffer around the place of residence. The study further shows that the aggregated benefits for well-being of a new 3.14 hectare park in an average German major city would add up to almost 934,000 € per year, an amount that would be four times higher than the potential costs for the construction and maintenance of this park.

Urban green space and real estate prices: Studies looking at the capitalization of urban green space in real estate prices try to value a bundle of ES connected with urban biodiversity ecosystems rather than a particular ES. An extensive body of literature applying the Hedonic Price Method exists at the international level, while a relatively low number of studies exist for China and Germany (e.g., Box 3.4).

However, the available studies in China and Germany provide an interesting view on social preferences for urban biodiversity and ES and their economic dimensions. A recent study from China explores the impacts of key environmental elements on the residential housing value of 652 dwelling units in the city of Guangzhou, including window orientation, green space view, floor height, proximity to wooded areas and water bodies, and exposure to traffic noise (Jim and Chen 2006). One of the main findings is that a view of green spaces and proximity to water bodies positively influence housing prices to an extent of 7.1 and 13.2%, respectively. The study demonstrates that HPM could be applied in the Chinese context with an increasingly expanding and privatized property market to inform policy and planning about the value of urban nature conservation and the design of ecological green space networks.

More recently, the capitalization of urban green in real estate prices has also been proven in the cities of Berlin and Cologne (Germany) (Kolbe and Wüstemann 2014; Wüstemann and Kolbe 2017a). In fact, an increase in green space coverage of 1% in a 500 m buffer around the real estate would lead to an average increase in house prices in Berlin of 410.20 € (0.13%). Therefore urban green space is capitalized in real estate prices to a significant amount, but the impact of most structural variables (e.g., size, number of rooms) is higher. The establishment of a new 3.14 hectare park in Cologne would lead to an increase in the benefits for the real estate directly benefiting from the park of about 1.27 million €. These benefits would outweigh the potential costs for the construction and maintenance of this park.

Box. 3.4 Residential property increase in Beijing (by Gaodi Xie):

Zhang et al. (2012b) surveyed the relationship between the average house prices of 76 residential areas and 14 parks in Beijing and measured the total benefits of 18,070 ha of public green spaces on residential property value by using the inventory data of urban green spaces (2009) and GIS techniques. Results showed that the values of residential property located 850–1604 m away from parks achieved a 0.5% to 14.1% increase in sales price. The overall benefit of Beijing's public green spaces for residential property was 2.86 billion RMB (RMB: Chinese currency, US$1 = RMB 6.83), and the average benefit per hectare of public green space was 0.16 million RMB, corresponding to 1.8–3.9 times the maintenance cost of green space in Beijing.

The economic benefits of public green spaces on residential property vary for different regions. From a regional perspective, the public green spaces in Chaoyang, Haidian, Xicheng, Dongcheng, and Fengtai contribute nearly 94% of the total benefit of property value increase. Specifically, the public green space in Chaoyang can achieve 928 million RMB from the increase of property value, followed by Haidian with 603 million RMB, Xicheng with 507 million RMB, Dongcheng with 400 million RMB and Fengtai with 261 million RMB. The residential properties in Changping and Shijingshan gain

72 million RMB and 68 million RMB due to public green spaces, respectively. The green spaces in other districts or counties provide only 2% of the total benefit. However, the average value increase per unit area of green space does not follow the same order. The highest value increase is for the public green space in Xicheng, followed by Dongcheng with 0.7 million RMB/ha. The next-highest value increase per unit area of green space occurs in Haidian, Chaoyang, and Fengtai. The remaining public green spaces in Beijing have a minor economic effect on property (Fig. 3.11).

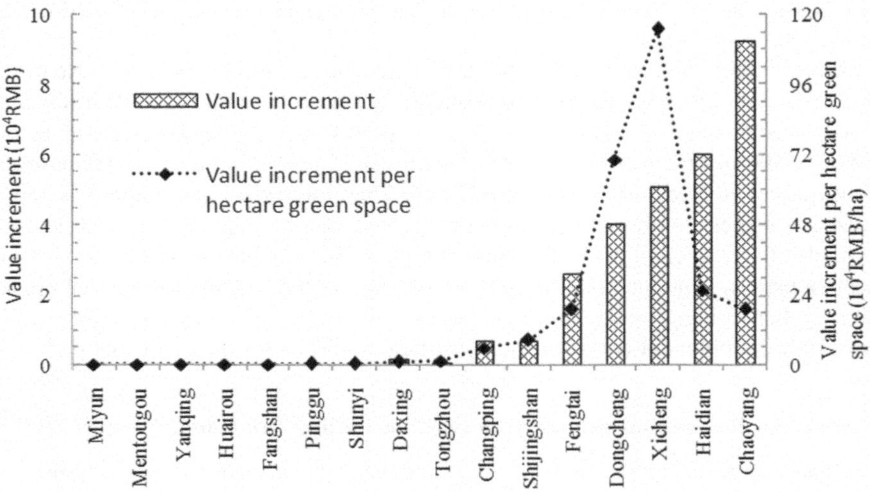

Fig. 3.11 Value increase of public green spaces in different regions of Beijing (modified after Zhang et al. 2012b)

Residents' perception of and attitudes toward green space: Several valuation studies in China and Germany provide data on social preferences for urban green space (Jim and Chen 2006; Chen and Jim 2011; Jim and Shan 2013; Bertram et al. 2017). An on-site survey of residents' perception of and attitudes toward urban green space in Guangzhou finds that stress reduction, health enhancement, children's development, and social interaction (to a lesser degree) are the main benefits associated with urban green. Socioeconomic variables significantly influence the perception, including gender, age, marital status, education, occupation, and district of residence (Jim and Shan 2013). In the process of sustainable urban planning, residents' positive attitudes toward urban green and appreciation of its benefits lend support to urban green space preservation and provision.

In Germany, a study by Bertram et al. (2017) examines recreational values of urban parks in Germany and identifies residents' social preferences for and attitudes toward urban park attributes. This study finds that on weekdays, respondents prefer urban parks with sports facilities in closer proximity to their homes, while the size of the parks is not so important. On the weekend, larger parks with picnic facilities are preferred, while distance matters less. Most important are, however, cleanliness and maintenance, regardless of whether a park is visited on weekdays or the weekend. The results underline the importance of considering different temporal contexts and variety-seeking behavior when preferences for outdoor recreation are concerned.

Costs and benefits of urban green space: Estimating the ES of additional green space in urban areas in monetary terms raises the question of the costs associated with the provision of urban green. Costs for additional green space mainly result from construction and maintenance of additional green space and from opportunity costs based on loss of profits mostly from real estate.

Zhang et al. (2012b) investigate maintenance costs of public green space in the city of Beijing and find that the increase in property tax revenue received by Beijing as a result of the enhanced value of properties around the public green spaces amounted to 2.86 billion RMB annually, the annual maintenance costs incurred by 18,070 ha of public green spaces are 0.7–1.6 billion RMB (according to the code for management of landscape greening in Beijing, the per year maintenance cost of grass space in the first category is 9 RMB m^{-2}, and the maintenance cost per year in the second and third category is 6 and 4 RMB m^{-2}, respectively), so the net annual income accruing to the city for its investment in the public green space is 1.2–2.1 billion RMB.

Krekel et al. (2016) show for the city of Berlin that the average construction and maintenance costs for one additional hectare of park in Berlin would lie—depending on park infrastructure and usage intensity—between 23,333 and 204,000 Euro annually. However, they also show that the benefits for residential well-being of this additional green space would exceed the construction and maintenance costs (Wüstemann and Kolbe 2017b).

Conclusion

Cities are facing enormous changes and challenges, including negative impacts resulting from climate and demographic change, high urbanization rates, and natural resource depletion. Research on the economic benefits of urban biodiversity and ES raises public awareness of the value of urban green and supports the discussion on sustainable urban development in both China and Germany. The findings demonstrate the effects of any land-use change on the provision of ES and provide decision makers with additional information in the process of urban planning. At the same time, valuation studies exhibit wide disparities in terms of methods and databases across the case studies. Main reasons are the lack of (i) accepted methods for quantification and valuation of the ES, (ii) standardization of data collection and analysis methods. Future research should focus on how to overcome these limitations and to integrate the multiple values of urban biodiversity into urban decision-making.

3.7 How Green Are Our Cities? Green Space Provision in Urban Areas

Ralf-Uwe Syrbe, Wei Hou, Karsten Grunewald and Juliane Mathey

As the previous chapters show, urban green spaces are multifunctional and bring various benefits for the quality of life in cities. But since these areas are in competition with other growing claims of land use, the essential questions are how many green spaces we need in our cities, of which quality they should be and how they can be secured in planning process (Sect. 3.4, 4.1 and Chap. 5).

Qualitative recommendations must complement quantitative benchmarks for urban green. Both are scale-dependent. Therefore, it has to be specified whether they concern an entire city (i.e., the entire urban green system within the administrative borders), one or more quarters of a city (district level) or an individual green space (site level). At each level, quantitative and qualitative aspects have to be taken into consideration. URGE-Team (2004) proposed several quantitative indicators for the city and site levels. The most important city- and district-level indicators of urban green space composition can be found in Table 3.8; they can be complemented by measures of configuration such as fragmentation vs. connectivity of green space. For the site level, some other indicators are available, such as size, shape, proportion of vegetation, which would exceed the scope of this chapter. Suitable qualitative criteria for the city and district level are: species and habitat diversity, level of protection (cultural and natural heritage), capacity of urban green to improve environmental quality (e.g., quality of air, soil and water, microclimate) and for the site level additionally: access, privacy and safety of the individual area (URGE-Team 2004).

This section focuses on quantitative indicators, thresholds and how to capture them for the city level. Before citywide benchmarks for the amount of green space can be defined, it has to be clarified how the amount and condition of green space can be measured. Here we present the main groups of indicators that can be clearly determined using widely available data and that can be associated with target values. The results will be compared for Chinese and German cities; for special findings on individual cities see Sect. 4.3.

Indicators and target values for green space in cities

The German National Strategy on Biological Diversity (BMU 2007) demands as a strategic aim for urban landscapes to considerably enhance the proportion of vegetated areas (green space here includes private gardens, green yards, overgrown walls and roofs etc., Sects. 1.2 and 2.2) in all settlements by 2020. Publicly accessible green space having manifold qualities and functions should in general be available within walking distance (BMUB 2015). To meet these aims, well communicable and clearly measurable target values are essential for urban planning and development. Precise specifications for the accessibility, quality, and maintenance of appropriate green space have to be promulgated. Strategies in China (Eco-city,

Table 3.8 Indicators for description and assessment of urban green space (drafted by the authors based on Dosch and Neubauer 2016; Grunewald et al. 2017a)

Indicator	Differentiation	Recording method	Target or orientation values
1. Green space proportion	• Type of green space (publicly accessible, relevant for recreation) • Delimitation of the city/urban area	Green space/total area in % 1. using topographical data 2. or using remote sensing (only total green recordable)	Examples of total green values: Beijing: 73.2%[a] Hamburg: 71.4%[b]
2. Green space per capita	• Type of green space (publicly accessible, relevant for recreation) • Delimitation of the city/urban area	Green space in m^2/number of inhabitants 1. using topographical data 2. or using remote sensing combined with population data (statistics)	for public green space only minimum: 9 m^2/inh. optimum: 50 m^2/inh. (WHO 2010) Germany: 15 m^2/inh. (DRL 2006)—22 m^2/inh. (Schröter 2015)
3. Green space accessibility	• Type of green space (public, recreational etc.) • Minimum size of green space considered • Maximum distance from inhabited area	Area of green space (of minimum size within given distance) in m^2, divided by the number of inhabitants who can access this green space	Germany, public green space[c]: 4 m^2/inh. > 0.2 ha—250 m 6 m^2/inh. > 5 ha—500 m 7 m^2/inh. > 10 ha—1000 m (+2 m^2/place of employment)
		Share of inhabitants with access to green space of specified size within given distance (in %)	• Scientific proposal: 100% • Political target in Germany: 95%[d]
4. Green space metrics	• Height of urban green • Volume of urban green • Tree canopy area	1. Average vegetation height (m) 2. Tree canopy proportion (%) 3. Green volume number (m^3/m^2) 4. Leaf area index (m^2/m^2)	Cf. Berlin Environmental Atlas (2013); Examples of tree canopy mean values: Beijing: 19.1%, Shanghai: 13.2%[e], Chengdu: 20.2%[f]
5. Soil sealing degree	• Regarding only traffic and built-up area • Regarding the density of pavement as well	Area of developed or paved surface divided by total area	Plot limits in BauNVO (2013): small housing 30% ordinary housing 60% special housing 80%

[a]Calculation by the authors
[b]Berliner Morgenpost (2016)
[c]GALK (1973)
[d]Grunewald et al. (2017a)
[e]Yang and Zhou (2009)
[f]See Sect. 4.3.7

Garden city; Sect. 2.2) point in the same direction. Green space standards are formalized within Chinese planning codes, but are difficult to enforce (Wolch et al. 2014).

Green space proportion indicators measure the share of vegetated area in comparison to the entire city, district, or site. If there are extended rural areas or larger forests within a reference area, the indicator value can be misleading, as the remote forests and fields create the impression of a large extent of green space in an actually densely built-up city. Furthermore, indicator values depend sensitively on what kind of green space (only public green or private as well, only recreationally relevant, or also agricultural land and wastelands) has been included in calculation. The latter applies also to the next group of indicators.

Green space per capita provides a better insight into the possible demand for green infrastructure by the inhabitants. The World Health Organization (WHO) demands at least 9 m^2/inh. green space in general and designates 50 m^2/inh. as preferable (WHO 2010). Schmidt et al. (2014) provided suggestions for the methodological framework and quality requirements as part of the WHO project "Health City Planning," which have been implemented for instance in Dresden and will be mentioned exemplarily below. The demand for green space is differentiated in a so-called basic requirement for public green (all kinds of parks and urban forest) and specific requirements for designated social groups such as children, elderly people (playgrounds, gardens, cemeteries, etc.).

Green space accessibility considers not only the available green space but also how to get there. It can be specified by linear distance or actual path length. These indicators are more reliable regarding city boundary effects, as green space at a larger distance is not considered, regardless of whether it belongs to a municipal area or not. Therefore, the indicator will not be affected by incorporation of villages and other types of changes in the city's area.

Green volume and tree canopy take the third dimension into account. Not only the area extension and local situation are important for the function of green space, but also the amount of biomass, height and stratification of plant stock as well as the leaf area can be measured with respect to their biophysical performance (e.g., relevant for microclimatic regulation effects; Lehmann et al. 2014).

As values complementary to the above, the **degree of soil sealing** as well as urban, building and population density play a role in the mass exchange (input of resources, output of wasted) and environmental situation of a city (Deilmann et al. 2017). Internationally, standard values for an optimal population density are under discussion, including justifications for upper and lower limits, which take into consideration in particular urban sprawl, sustainable land use, and densification (Jaeger et al. 2010). Values of population density are often mentioned that range from 3890 inh./km^2 in Berlin to 4329 inh./km^2 in Shanghai. These values do not give an idea of the amount of green space supplied.

Target values in Germany were set already 100 years ago and updated by the conference of garden office directors (GALK 1973). For instance, 20 m^2 of public green space per capita is demanded there. This general value includes 6 m^2/inh. of smaller parks of size at least 0.5 ha at a maximal walking distance of 500 m and

7 m^2/inh. of parks with more than 10 ha and at 1000 m distance. Other GALK standard values concern cemeteries (5 m^2/inh.), allotment gardens (10–12 m^2/inh.), sports fields (3.5 m^2/inh.), and open-air pools (1 m^2/inh.). In Germany, 39% of the cities currently use benchmarks to measure green space (Kühnau et al. 2016). The German land-use ordinance (BauNVO 2013) places several limits on the construction density on individual parcels, depending on their classification in the land-use plan. For instance, ordinary housing plots should comply with a site occupancy index of 40% and not exceed a maximum of 60%, which also applies to the soil sealing degree (Table 3.8).

In China, target values exist mainly at the city level, but only a part of them are obligatory. Guiyang City (2008), for instance, specified the forest coverage (45%), others fixed standards for green area per capita (Guiyang City > 10 m^2, Tanjin > 12 m^2, Caofeidian > 20 m^2) or green area share in total area (Caofeidian 35%). In most cities, urban planning defines specific targets and standard values depending on the local situation (Zhou et al. 2012). The city of Xuzhou, for instance, has set precise values for the composition of green space (41% grassland and 44% forest area) and aims to provide the population with 13.6 m^2 green space per capita (already achieved—see Fig. 4.33). The city's green infrastructure plan aims to provide 4279 ha of park and similar green space, 27 ha of productive green space and 3821 ha of green distant space with new projects in Xuzhou, which are applied extensively due to the termination of coal mining in this area.

Indicator values for urban green space provision in Germany and China

Grunewald et al. (2017a) calculated the green space provision per inhabitant related to the total amount of green space and the accessibility of green space for all major German cities (Sect. 3.4). Table 3.9 summarizes the results for selected cities and compares them with other values from both countries. Larger cities have more problems in ensuring the provision and accessibility of green space. Results show that in general, cities with less inhabitants can provide more green space than larger cities with a higher population and a corresponding pressure on open space for development. However, there is no statistically significant correlation between indicator values and size of municipal area or between indicator values and size of population (Grunewald et al. 2017a).

Studies in Shanghai found that many residents lack access to parks, and that entire areas of the city have no formal green space (Yin and Xu 2009).

The proportion of urban green space has just recently been mapped as an indicator for instance by the newspaper "Berliner Morgenpost" (2016) using satellite data (total values for German cities in Table 3.9). These values are less precise than terrestrially determined data due to the uncertainty of remote sensing analysis in general. They include each green space that is recognizable from space, regardless of its accessibility, quality, or size. The values belong to the indicator "Green space proportion," recording method 2, and are roughly comparable with the first row of Table 3.8. To be sure, rating systems tend to assign numerical grades to things that are partially or entirely subjective; and of course green space coverage can be calculated in a variety of ways, leading to different results or

Table 3.9 Provision and accessibility of green space for selected major cities in China and Germany

City	Inh. (Mio.)/ area (km^2)[a]	Green-space proportion (%) recreational/total	Green-space per capita (m^2/inh.) recreational/total	Green-space accessibility (% of inh.)	Soil sealing degree (%)
Nanjing	5.0[a]/968[b]	8/38[c]	4.27/24.25[c]	35[c]	52[c]
Suzhou	2.0[a]/794[b]	6/46[c]	4.76/50.24[c]	17[c]	51[c]
Xuzhou	1.04[a]/217[b]	11/33[c]	6.64/25.86[c]	30[c]	62[c]
Beijing	11.3[a]/109[b]	11/37[c]	3.59/14.75[c]	33[c]	63[c]
Shijiazhuang	2.2[a]/258[b]	1/17[c]	1.02/11.15[c]	9[c]	74[c]
Tianjin	4.2[a]/337[b]	2/30[c]	1.25/18.13[c]	31[c]	68[c]
Berlin	3.5/892[d]	32[f]/59[e]	88[f]/150[d]	61[f]	39[c]
Hamburg	1.8/755[d]	36[f]/71[e]	159[f]/310[d]	70[f]	36[g]
Munich	1.4/311[d]	20[f]/49[e]	46[f]/110[d]	63[f]	47[g]
Cologne	1.0/405[d]	32[f]/58[e]	127[f]/230[d]	74[f]	34[g]
Leipzig	0.5/297[d]	21[f]/42[e]	125[f]/230[d]	61[f]	29[g]
Dresden	0.5/328[d]	39[f]/69[e]	251[f]/420[d]	60[f]	29[g]
Bonn	0.3/141[d]	25[f]/72[e]	107[f]/320[d]	76[f]	30[g]

Data sources
[a]population data from the City Statistics Yearbook 2015
[b]urban center area, not the city administrative area
[c]calculation by the authors based on the urban center area
[d]see Table 2.7
[e]Berliner Morgenpost (2016)
[f]calculation by the authors, for methods see Grunewald et al. (2017a)
[g]http://www.ioer-monitor.de/en/home/ (2012, for Leipzig: 2009)

rankings (Fuller and Gaston 2009). Everything depends on definitions (e.g., is the number related to municipal boundaries or to actual patterns of settlements). City rankings can be problematic, but they can also be useful if they get us talking about what is really important.

Kabisch et al. (2016) showed that the share of the population in European cities living within 500 m linear distance from green and forest areas with a minimum size of 2 ha ranges from 11 to 98%. For the city of Berlin, Germany, they found that 30% of the population lives within 300 m and 68% within 500 m distance. Additionally, they found that on the basis of official municipal data 58.7% of the population of Berlin has access to urban green space within 300 m distance. The latter value is slightly below our result for Berlin (Table 3.9). Possible reasons for a higher indicator value might include a broader definition of green space relevant to recreation (additionally considered: water area, meadow-orchard, cemetery) as well as the inclusion of green space outside the municipality under consideration.

In the French city of Nantes, which won the title of European Green Capital in 2013, 100% of the population lives within 300 m from green space. The environmental assessment of the European Environment Agency showed significant differences in green space provision between European cities ranging from Brussels, Copenhagen, and Paris, where all citizens live within 15 min walking distance from public green, to Venice and Kiev, where the corresponding figure is 63 and 47% of the population, respectively (Stanners and Bourdeau 1995).

Green space provision has often been expressed in terms of size per inhabitant, which is not really sufficient (large variation between different parts of the settlement, strongly influenced by where the outer border is drawn). The combination of the sizes of green space with distances allows for relatively simple modeling of the accessibility of green space and represents an effective methodology for a nationwide application. Deficits and trends can be pointed out, and comparisons between cities are possible, also internationally if the methodology is comparable. However, such nationwide calculations can only give an overview. Therefore, it is possible for cities which have higher-resolution data available to carry out more complex analyses in order to underpin the determined quantitative values by respective site-specific quality requirements for municipal planning of green space.

A GIS-based calculation (e.g., based on biotope mappings) of specific indicator values for city districts enables the identification of spatial need for action in every city and gives the municipal authority a framework for targeted maintenance and improvement of green space at the specific planning levels (e.g., Lehmann et al. 2014; Schmidt et al. 2014; Grunewald et al. 2017b).

Figures 3.12 and 3.13 show district-specific values for green space provision and soil sealing (pavement) in German and Chinese cities. Strong differences become apparent here, for instance regarding Beijing, where the green space proportion ranges from 5.1% in the city center to 100% in the mountainous areas, whereas distant green area will not satisfy all needs of the people in the center of the metropolis. Xuzhou offers green space also in proximity to the city, giving its people good opportunities for recreation after work and on weekends, also saving high costs and emissions for the traffic to go there. In comparison, the two German cities Berlin and Dresden show green space also in some populated areas near the center, but the differences within the two cities are very high. Note the different scales of the two maps!

Probably, a green space provision indicator based on distance can be more readily interpreted, but it is more difficult to assess since it requires fine-scaled population data. Therefore, soil sealing has been added on the right side of each figure. This proportion of developed and paved areas provides a kind of inverse number to the green space proportion. Sealed areas cannot percolate water and no longer have viable natural soils, and thus they are hardly suitable for outdoor recreation and have low potentials for biodiversity. The maps show clearly the densest built-up areas of the cities which are not only in the centers. These areas give reason to demand action to provide more green of any possible type, for instance through roof greening and green façades (see Sects. 4.2 and 4.3).

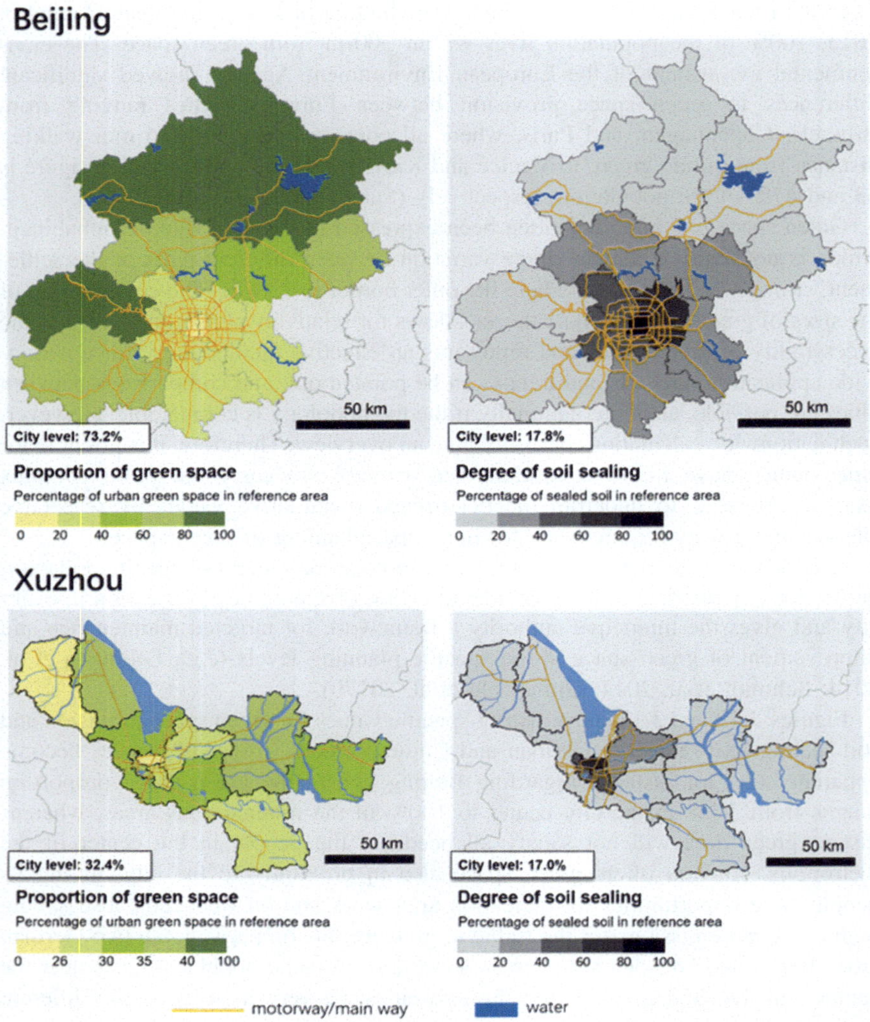

Fig. 3.12 Beijing and Xuzhou (China): proportion of green space (*left*) and degree of soil sealing (*right*). © W. Hou

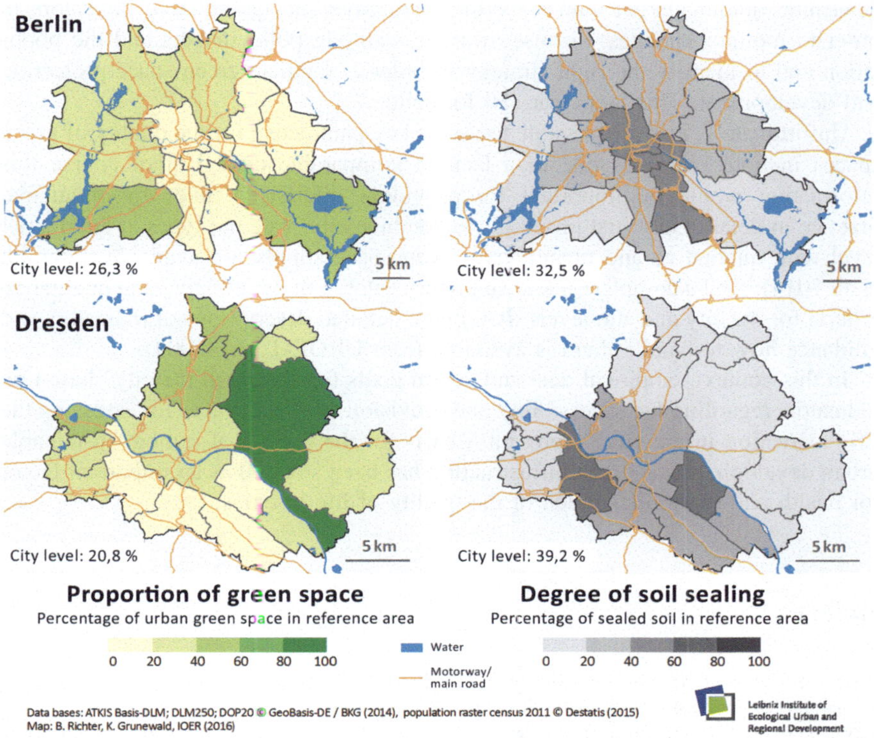

Fig. 3.13 Berlin and Dresden (Germany): proportion of green space (*left*) and degree of soil sealing (*right*). © B. Richter; K. Grunewald

Conclusion

A comparison with the standard and target values shows that

- the target of 9 m^2 green space per capita (WHO) or 20 m^2 (GALK 1973) is exceeded in all investigated case study cities in Germany (Table 3.9);
- by contrast, the target to provide 95% of the inhabitants with access to green space at walking distance (suggestion for Germany, Grunewald et al. 2017a) is missed (Table 3.9).

The indicator values for German cities, calculated with a comprehensive methodology, are updatable, but the recent trend (slight decrease) is still uncertain. However, first systematic trend analyses using the example of selected cities (Dresden, Shanghai, Guangzhou) prove that the efforts toward more green space were successful in the last five years, as the proportion of urban parks and forests increased (see Sect. 4.3.5) at the expense of farmland and unused land.

However, the population growth in the biggest cities of China and Germany can endanger the provision of green space in the near future. Therefore, it is essential to determine binding target values for the qualitative and quantitative indicators in order to monitor changes, to raise awareness among policymakers and the population and to identify the right strategy in order to regulate green space protection and development. This is a clear call for action.

Unfortunately, it is not possible to assess key parameters for the quality of green spaces in both countries due to a lack of comparable data. But of course this information would be important for providing reference values for municipal practice in detail. Only first proposals to combine city structure types at the district level with qualitative and quantitative green space supply are available (Schmidt et al. 2014). And a complete interdisciplinary catalog of quantitative and qualitative criteria for the city and site levels (ICC) with detailed descriptions of indicators and guidance how to handle them is available from URGE-Team (2004).

In this context, empirical data and action goals for "green in the city" based on indicators regarding the accessibility and provision of green spaces for assessing the ES "recreation in the city" represent a basis for the pursuit of a more sustainable urban development, as green infrastructure has been stressed as an important factor for health and constitutes much of the quality of life in cities.

References

Abkar M, Kamal M, Mariapan M, Maulan S, Sheybani M (2010) The role of urban green spaces in mood change. Aust J Basic Appl Sci 4(10):5352–5361

Aevermann T, Schmude J (2015) Quantification and monetary valuation or urban ecosystem services in Munich, Germany. Zeitschrift für Wirtschaftsgeographie 59(3):188–200

Akbari H, Pomerantz M, Taha H (2001) Cool surfaces and shade trees to reduce energy use and improve air quality in urban areas. Sol Energy 70(3):295–310

Alcock I, White MP, Wheeler BW, Fleming LE, Depledge MH (2014) Longitudinal effects on mental health of moving to greener and less green urban areas. Environ Sci Technol 48 (2):1247–1255

Alriksson S, Öberg T (2008) Conjoint analysis for environmental evaluation: a review of methods and applications. Environ Sci Pollut Res 15:244–257

Andersson E, Tengö M, McPhearson T, Kremer P (2015) Cultural ecosystem services as a gateway for improving urban sustainability. Ecosyst Serv 12:165–168

Annerstedt M, Ostergren PO, Björk J, Grahn P, Skärbäck E, Währborg G (2012) Green qualities in the neighbourhood and mental health: results from a longitudinal cohort study in Southern Sweden. BMC Public Health 12:337

Arnfield AJ (2003) Two decades of urban climate research: a review of turbulence, exchanges of energy and water, and the urban heat island. Int J Climatol 23:1–26

Artmann M (2013) Spatial dimensions of soil sealing management in growing and shrinking cities—a systemic multi-scale analysis in Germany. Erdkunde 67(3):249–264

Barbosa O (2007) Who benefits from access to green space? A case study from Sheffield, UK. Landscape Urban Plann 83:187–195

Baro F, Chaparro L, Gomez-Baggethun E, Langemeyer J, Nowak DJ, Terradas J (2014) Contribution of ecosystem services to air quality and climate change mitigation policies: the case of urban forests in Barcelona, Spain. Ambio 43:466–479

Barradas V (1991) Air temperature and humidity and human comfort index of some city parks of Mexico City. Int J Biometeorol 35:24–28

Bastian O, Schreiber KF (eds) (1994) Analyse und ökologische Bewertung der Landschaft. G. Fischer-Verlag, Jena, Stuttgart, 502 pp. 2nd, considerably revised edn. 1999: Spektrum Akadem. Verlag, Heidelberg, Berlin, 564 pp.

Bastian O, Steinhardt U (eds) (2002) Development and perspectives in landscape ecology. Kluwer Academic Publishers, Dordrecht, 527 pp.

Bateman I (1993) Evaluation of the environment: a survey of revealed preference techniques. Tech. rept, GEC

Bateman IJ, Carson RT, Day BH, Hanemann WM, Hanley N, Hett T et al (2002) Economic valuation with stated preferences techniques: a manual. Edward Elgar, Cheltenham

BauNVO—Verordnung über die bauliche Nutzung der Grundstücke (2013) Baunutzungsverordnung in the version proclaimed on 23 Januar 1990 (BGBl. I p. 132), last amended by Article 2 of the law of 11 June 2013 (BGBl. I S. 1548)

Berlin Environmental Atlas (2013) http://www.stadtentwicklung.berlin.de/umwelt/umweltatlas/edua_index.shtml. Accessed 14 June 2017

Berliner Morgenpost (2016) Das sind Deutschlands grünste Städte. Viele Städte behaupten von sich, besonders viele Grünflächen zu bieten. Die Berliner Morgenpost hat Satellitenbilder ausgewertet und zeigt erstmals, wie grün Deutschland wirklich ist. http://interaktiv.morgenpost.de/gruenste-staedte-deutschlands/ Accessed 14 Sept 2016

Bertram C, Rehdanz K (2015) The role of urban green space for human well-being. Ecol Econ 120:139–152

Bertram C, Meyerhoff J, Rehdanz K, Wüstemann H (2017) Differences in the recreational value of urban parks between weekdays and weekends: a discrete choice analysis. Landscape Urban Plann 159:5–14

BfN—Bundesamt für Naturschutz (2009) Biological diversity and cities: a review and bibliography. In: Werner P, Zahner R (eds) BfN-Skripten 245

Birmili W, Hoffmann T (2005) Particulate and dust pollution, inorganic and organic compounds. In: Encyclopedia of environmental pollutants. Elsevier Ltd

Blume H-P, Sukopp H (1976) Ökologische Bedeutung anthropogener Bodenveränderungen. Schr.-R. f Vegetationskunde 10:75–89

BMU—Bundesministerium für Umwelt, Naturschutz und Reaktorsicherheit (ed) (2007) Nationale Strategie zur biologischen Vielfalt. Bonn. 178 pp.

BMUB—Bundesministerium für Umwelt, Naturschutz, Bau und Reaktorsicherheit (2015) Grün in der Stadt – Für eine lebenswerte Zukunft. Grünbuch Stadtgrün (Green Book „Green in the City“). Federal Ministry for the Environment, Nature Conservation, Building and Nuclear Safety, Berlin. http://www.bmub.bund.de/service/publikationen/downloads/details/artikel/gruen-in-der-stadt-fuer-eine-lebenswerte-zukunft/?tx_ttnews[backPid]=289. Accessed 27 June 2015

Bolund P, Hunhammar S (1999) Ecosystem services in urban areas. Ecol Econ 29(2):293–301

Bongardt B (2006) Stadtklimatologische Bedeutung kleiner Parkanlagen—dargestellt am Beispiel des Dortmunder Westparks. (Universität Duisburg-Essen, Dissertation), Hohenwarsleben (Westarp Wissenschaften), Essener Ökologische Schriften 24, 268 pp

Bowler DE, Buyung-Ali L, Knight TM, Pullin AS (2010a) Urban greening to cool towns and cities: a systematic review of the empirical evidence. Landscape Urban Plann 97:147–155

Bowler DE, Lisette MB-A, Knight TM, Pullin AS (2010b) A systematic review of evidence for the added benefits to health of exposure to natural environments. BMC Public Health 10:456

Breuste J, Qureshi S, Li J (2013a) Scaling down the ecosystem services at local level for urban parks of three megacities. Hercynia N. F. 46:1–20

Breuste J, Schnellinger J, Qureshi S, Faggi A (2013b) Urban ecosystem services on the local level: urban green spaces as providers. Ekologia 32(3):290–304

Bruse M, Fleer H (1998) Simulating surface-plant-air interactions inside urban environments with a three dimensional numerical model. Environ Model Softw 13:373–384

Bullinger M, Alonso J, Apolone G et al (1998) Translating health status questionnaires and evaluating their quality: the IQOLA project approach. Clin Epidemiol 51:913–923

Bundesregierung (2008) Deutsche Anpassungsstrategie an den Klimawandel (DAS), Berlin, 78 pp

Burkart K, Canário P, Scherber K, Breitner S, Schneider A, Alcoforado MJ, Endlicher W (2013) Interactive short-term effects of equivalent temperature and air pollution on human mortality in Berlin and Lisbon. Environ Pollut 183:54–63
Calfapietra C, Morani A, Sgrigna G et al (2016) Removal of ozone by urban and peri-urban forests: evidence from laboratory, field, and modeling approaches. J Environ Qual 45:224–233
Chai YX, Zhu N, Han HJ (2002) Dust removal of urban tree species in Harbin. Chin J Appl Ecol 13(9):1121–1126
Chang C-R, Li M-H, Chang S-D (2007) A preliminary study on the local cool-island intensity of Taipei city parks. Landscape Urban Plann 80:386–395
Chen J (2007) Rapid urbanization in China: a real challenge to soil protection and food security. Catena 69:1–15
Chen WY (2015) The role of urban green infrastructure in offsetting carbon emissions in 35 major Chinese cities: a nationwide estimate. Cities 44:112–120
Chen W, He XY, Zhang Y et al (2003) Dust absorption effect of urban conifers in Northeast China. Chin J Appl Ecol 14(12):2113–2116
Chen WY, Jim CY (2011) Resident valuation and expectation of the urban greening project in Zhuhai, China. J Environ Planning Manage 54(7):851–869
Chen X-L, Zhao H-M, Li P-X, Yin Z-Y (2006) Remote sensing image-based analysis of the relationship between urban heat island and land use/cover changes. Remote Sens Environ 104:133–146
Cheng X, Wei B, Chen G, Li J, Song C (2015) Influence of park size and its surrounding urban landscape patterns on the park cooling effect. J Urban Plann Dev 141(3):A4014002
Chou S-Z, Zhang C (1982) On the Shanghai urban heat island effect. Acta Geogr Sin 37:372–382
COM (2011) Final communication from the commission to the European Parliament, the Council, the Economic and Social Committee and the Committee of the Regions: our life insurance, our natural capital: an EU biodiversity strategy to 2020. European Commission, Brussels 3 May 2011
Comber A, Brunsdon C, Green E (2008) Using a GIS-based network analysis to determine urban greenspace accessibility for different ethnic and religious groups. Landscape Urban Plann 86:103–114
Coon JT, Boddy K, Stein K, Whear R, Barton J, Depledge MH (2011) Does participating in physical activity in outdoor natural environments have a greater effect on physical and mental well-being than physical activity indoors? A systematic review. Environ Sci Technol 45 (5):1761–1772
Cowell FR (1978) The garden as a fine art: from antiquity to modern times. Joseph, London
Currie BA, Bass B (2008) Estimates of air pollution mitigation with green plants and green roofs using the UFORE model. Urban Ecosyst 11:409–422
Deilmann C, Lehmann I, Schumacher U, Behnisch M (eds) (2017) Stadt im Spannungsfeld von Kompaktheit, Effizienz und Umweltqualität. Anwendungen urbaner Metrik. Springer, Heidelberg
Derkzen ML, van Teeffelen AJA, Verburg PH (2015) Quantifying urban ecosystem services based on high-resolution data of urban green space: an assessment for Rotterdam, the Netherlands. J Appl Ecol. doi:10.1111/1365-2664.12469
de Vries S, Verheij RA, Groenewegen PP, Spreeuwenberg P (2003) Natural environments—healthy environments? An exploratory analysis of the relationship between green space and health. Environ Plann 35(10):1717–1731
Dieter M, Elsasser P (2002) Carbon stocks and carbon stock changes in the tree biomass of Germany's forests. Forstwissenschaftliches Centralblatt 121(4):195–210
Dong N (2006) Shanghais innerstädtischer Freiraumwandel in zehn Jahren Stadterneuerung von 1991–2000 anhand von Beispielen aus Huangpu, Nanshi, Luwan, Jing'an und Lujiazui. Ph.D. thesis, University Kassel
Dong Z, Wang J (2011) Quantitative standard of eco-compensation for the water source area in the middle route of the South-to-North Water Transfer Project in China. Front Environ Sci Eng China 5(3):459–473

Dosch F, Neubauer U (2016) Kennwerte für grüne Infrastruktur. Sicherung städtischer Freiraumqualität durch Richt- und Orientierungswerte? RaumPlanung 185:15–39

DRL—Deutscher Rat für Landespflege (2006). Durch doppelte Innenentwicklung Freiraumqualitäten erhalten. Schriftenreihe des Deutschen Rates für Landespflege 78:5–39

Easterling DR, Evans JL, Groisman PY, Karl TR, Kunkel KE, Ambenje P (2000) Observed variability and trends in extreme climate events: a brief review. Bull Am Meteor Soc 81:417–425

EEA—European Environment Agency (2013) Air quality in Europe—2013 report. Publications Office of the European Union, Luxembourg

Ellert U, Kurth B-M (2004) Methodische Betrachtungen zu den Summenscores des SF-36 anhand der erwachsenen Bevölkerung. Bundesgesundheitsbl-Gesundheitsforsch-Gesundheitsschutz 47:1027–1032

Elmqvist T, Setälä H, Handel SN et al (2015) Benefits of restoring ecosystem services in urban areas. Curr Opin Environ Sci 14:101–108

Farivar SS, Cunningham WE, Hays RD (2007) Correlated physical and mental health summary scores for the SF-36 and SF-12 Health Survey. V.I. Health Qual Life Outcomes 5:54

Feng CY, Gao JX, Tian MR et al (2007) Research on dust absorption ability and efficiency of natural vegetation in Mentougou District, Beijing. Res Environ Sci 20(5):155–159

Frich P, Alexander L, Della-Marta P, Gleason B, Haylock M, Klein Tank A, Peterson T (2002) Observed coherent changes in climatic extremes during the second half of the twentieth century. Climate Res 19 193–212

Fukuhara S, Bito S, Green J et al (1998) Translation, adaption, and validation of the SF-36 Health Survey for use in Japan. J Clin Epidemiol 51:1037–1044

Fuller RA, Gaston KJ (2009) The scaling of green space coverage in European cities. Biol Lett. doi:10.1098/rsbl.2009.0010

GALK—Gartenamtsleiterkonferenz (1973) Richtwerte der ständigen Konferenz der Gartenamtsleiter

Garrod G, Willis KG (1999) Economic valuation of the environment. Edward Elgar, Cheltenham

Gill SE, Handley JF, Ennos AR, Pauleit S (2007) Adapting cities for climate change: the role of the green infrastructure. Built Environ 33(1):115–133

Gómez-Baggethun E, Gren Å, Barton DN, Langemeyer J, McPhearson T, O'Farrell P, Andersson E, Hamstead Z, Kremer P (2013) Urban Ecosystem Services. In: Elmqvist T et al (eds) Urbanization, biodiversity and ecosystem services: challenges and opportunities. A global assessment. A Part of the Cities and Biodiversity Outlook Project. Springer, Netherlands, pp 175–251

GOOOOD (2014) Ningbo Eco Corridor. http://www.gooood.hk/ningbo-eco-corridor-by-swa.htm?lang=en_US, Accessed 10 Mar 2014

Grahn P, Stigsdotter UA (2003) Landscape planning and stress. Urban Forest Urban Greening 2 (1):1–18

Grewal SS, Grewal PS (2012) Can cities become self-reliant in food? Cities 29:1–11

Grunewald K, Bastian O (eds) (2015) Ecosystem services—concept, methods and case studies. Springer, Berlin

Grunewald K, Richter B, Meinel G, Herold H, Syrbe R-U (2017a) Proposal of indicators regarding the provision and accessibility of green spaces for assessing the ecosystem service "recreation in the city" in Germany. Int J Biodivers Sci Ecosyst Serv Manage 13(2):26–39. doi:10.1080/21513732.2017.1283361

Grunewald K, Richter B, Behnisch M (2017b) Urban green space indicators at the city and city-district level—calculated and discussed for German cities. Ecol Ind (in review)

Haase D (2009) Effects of urbanisation on the water balance—a long-term trajectory. Environ Impact Assess Rev 29(4):211–219

Haase D, Larondelle N, Andersson E, Artmann M, Borgström S, Breuste J et al (2014) A quantitative review of urban ecosystem service assessments: concepts, models, and implementation. Ambio 43(4):413–433

Haines-Young R, Potschin M (2013) Common International Classification of Ecosystem Services (CICES). Report to the European Environmental Agency, EEA Framework, Contract EEA/IEA/09/003

Harrison PA, Berry PM, Simpson S, Haslett JR, Blicharska M et al (2014) Linkages between biodiversity attributes and ecosystem services: a systematic review. Ecosyst Serv 9:191–203

Hartig T, Evans GW, Jamner LD, Davis DS, Garling T (2003) Tracking restoration in natural and urban settings. J Environ Psychol 23(2):109–123

Hartje V, Wüstemann H, Bonn A (2015) Naturkapital und Klimapolitik—Synergien und Konflikte. Naturkapital Deutschland—TEEB DE report (Natural capital and climate politics: synergies and conflicts. The Economics of Ecosystems and Biodiversity—Germany report). Technische Universität Berlin, Berlin; Helmholtz-Zentrum für Umweltforschung—UFZ, Leipzig

Hernández-Morcillo M, Plieninger T, Bieling C (2013) An empirical review of cultural ecosystem service indicators. Ecol Ind 29:434–444

Hlatky MA, Boothroyd D, Vittinghoff E et al (2002) Quality-of-life and depressive symptoms in postmenopausal women after receiving hormone therapy. JAMA 287:591–597

Höppe P (1999) The physiological equivalent temperature—a universal index for the biometeorological assessment of the thermal environment. Int J Biometeorol 43(2):71–75

Honjo T, Takakura T (1990) Simulation of thermal effects of urban green areas on their surrounding areas. Energy Build 15:443–446

Honold J, Lakes T, Beyer R, van der Meer E (2015) Restoration in urban spaces: nature views from home, greenways, and public parks. Environ Behav 1–30

Huang YJ, Akbari H, Taha H, Rosenfeld AH (1987) The potential of vegetation in reducing summer cooling loads in residential buildings. J Climate Appl Meteorol 26:1103–1116

Illgen M (2011) Hydrology of urban environments. In: Niemelä J (ed) Urban ecology: patterns, processes, and application. Oxford University Press, New York, pp 59–70

IPCC—Intergovernmental Panel on Climate Change (2013) Summary for policymakers. In: Climate change 2013: the physical science basis. Contribution of working group I to the fifth assessment report of the Intergovernmental Panel on Climate Change. In: Stocker TF, Qin D, Plattner G-K, Tignor M, Allen SK, Boschung J, Nauels A, Xia Y, Bex V, Midgley PM (eds) Cambridge University Press, Cambridge, United Kingdom and New York, NY, USA

Jaeger JAG, Bertiller R, Schwick C, Kienast F (2010) Suitability criteria for measures of urban sprawl. Ecol Ind 10(2):397–406

Jauregui E (1990) Influence of a large urban park on temperature and convective precipitation in a tropical city. Energy Build 15:457–463

Jendritzky G, Bröde P, Fiala D, Havenith G, Weihs P, Batcherova E, De Dear R (2009) Der Thermische Klimaindex UTCI. In: Deutscher Wetterdienst (ed) Klimastatusbericht 2009. Offenbach a.M., pp. 96–101

Jim CY, Chen WY (2006) Recreation—amenity use and contingent valuation of urban greenspaces in Guangzhou, China. Landscape Urban Plann 75(1, 2):81–96

Jim CY, Chen WY (2009) Ecosystem services and valuation of urban forests in China. Cities 26 (4):187–194

Jim CY, Shan X (2013) Socioeconomic effect on perception of urban green spaces in Guangzhou, China. Cities 31:123–131

Judge TA, Watanabe S (1993) Another look at the job satisfaction-life satisfaction relationship. J Appl Psychol 78(6):939–948

Jusuf SK, Wong NH, Hagen E, Anggoro R, Hong Y (2007) The influence of land use on the urban heat island in Singapore. Habitat Int 31:232–242

Kabisch N, Haase D (2013) Green spaces of European cities revisited for 1990–2006. Landscape Urban Plann 110:113–122

Kabisch N, Haase D (2014) Green justice or just green? Provision of urban green spaces in Berlin, Germany. Landscape Urban Plann 122:129–139

Kabisch N, Strohbach M, Haase D, Kronenberg J (2016) Urban green space availability in European cities. Ecol Ind. doi:10.1016/j.ecolind.2016.02.029

Kaczynski AT, Henderson KA (2007) Environmental correlates of physical activity: a review of evidence about parks and recreation. Leisure Sci 29(4):315–354

Kahneman D, Sugden R (2005) Experienced utility as a standard of policy evaluation. Environ Resource Econ 32:161–181

Kallweit D, Wintermeyer D (2013) Berechnung der gesundheitlichen Belastung der Bevölkerung in Deutschland durch Feinstaub (PM10). UMID: Umwelt und Mensch – Informationsdienst 4:18–24

Knecht C (2004) Urban nature and well-being: some empirical support and design implications. Berkeley Plann J 17(1):82–108

Kolbe J, Wüstemann H (2014) Estimating the value of urban green space: a hedonic pricing analysis of the housing market in Cologne, Germany. Folia Oeconomica 5(307):45–61

Kowarik I (2005) Wild urban woodlands: towards a conceptual framework. In: Kowarik I, Körner S (eds) Wild urban woodlands. New perspectives for urban forestry. Springer, Heidelberg, pp 1–32

Krekel C, Kolbe J, Wüstemann H (2015) The greener, the happier? The effects of urban green and abandoned areas on residential well-being. SOEPpapers 728

Krekel C, Kolbe J, Wüstemann H (2016) The greener, the happier? The effect of urban land use on residential well-being. Ecol Econ 121:117–127

Kremer P, Hamstead ZA, McPhearson T (2016) The value of urban ecosystem services in New York City: A spatially explicit multicriteria analysis of landscape scale valuation scenarios. Environ Sci Policy 62:57–68

Kroll F, Müller F, Haase D, Fohrer N (2012) Rural–urban gradient analysis of ecosystem services supply and demand dynamics. Land Use Policy 29:521–535

Krüger T, Held F, Hoechstetter S (2014) Identifikation von hitzesensitiven Stadtquartieren. In: Wende W, Rößler S, Krüger T (eds) Grundlagen für eine klimawandelangepasste Stadt- und Freiraumplanung. REGKLAM (Regionales Klimaanpassungsprogramm Modellregion Dresden), Issue 6. Rhombos, Berlin, pp 5–20

Kühnau C, Böhme C, Bunzel A, Böhm J, Reinke M (2016) Von der Theorie zur Umsetzung: Stadtnatur und doppelte Innenentwicklung (From theory into practice: urban green space and dual inner development). Natur und Landschaft 7:329–335

Kuo FE, Bacaicoa M, Sullivan WC (1998) Transforming inner city landscapes: trees, sense of place, and preference. Environ Behav 30(1):28–59

Kuo M (2015) How might contact with nature promote human health? Promising mechanisms and a possible central pathway. Front Psychol 6:1093

Kuttler W, Weber S, Schonnefeld J, Hesselschwerdt A (2007) Urban/rural atmospheric water vapour pressure differences and urban moisture excess in Krefeld, Germany. Int J Climatol 27:2005–2015

Larsen L (2015) Urban climate and adaptation strategies. Front Ecol Environ 13:486–492

Lee AC, Jordan HC, Horsley J (2015) Value of urban green spaces in promoting healthy living and wellbeing: prospects for planning. Risk Manage Health Policy 8:131–137

Lehmann I, Mathey J, Rößler S, Bräuer A, Goldberg V (2014) Urban vegetation structure types as a methodological approach for identifying ecosystem services—application to the analysis of micro-climatic effects. Ecol Ind 42:58–72

Leiva GMA, Santibañez DA, Ibarra ES, Matus CP, Seguel R (2013) A five-year study of particulate matter (PM2.5) and cerebrovascular diseases. Environ Pollut 181:1–6

Leng PS, Yang XH, Su F, Wu B (2004) Economic valuation of urban greenspace ecological benefits in Beijing City. J Beijing Agric Coll 19(4):25–28

Li J, Song C, Cao L, Zhu F, Meng X, Wu J (2011) Impacts of landscape structure on surface urban heat islands: a case study of Shanghai, China. Remote Sens Environ 115:3249–3263

Li X, Zhou W, Ouyang Z, Xu W, Zheng H (2012) Spatial pattern of greenspace affects land surface temperature: evidence from the heavily urbanized Beijing metropolitan area, China. Landscape Ecol 27:887–898

Lupp G, Förster B, Kantelberg VM, Arkmann T, Naumann J, Honert C, Koch M, Pauleit S (2016) Assessing the recreation value of urban woodland using the Ecosystem Service Approach in two forests in the Munich Metropolitan Region. Sustainability 8(1156):1–14

Lu XC, Yao T, Fung JCH et al (2016) Estimation of health and economic costs of air pollution over the Pearl River Delta region in China. Sci Total Environ 566–567:134–143

Maas J, Verheij RA, Groenewegen PP, de Vries S, Spreeuwenberg P (2006) Green space, urbanity, and health: how strong is the relation? J Epidemiol Community Health 60 (7):587–592

Maas J, Verheij RA, Spreeuwenberg P, Groenewegen PP (2008) Physical activity as a possible mechanism behind the relationship between green space and health: a multilevel analysis. BMC Public Health 8:260–273

Maas J, Verheij RA, de Vries S, Spreeuwenberg P, Schellevis FG, Groenewegen PP (2009) Morbidity is related to a green living environment. J Epidemiol Community Health 63 (12):967–973

Maraja R, Barkmann J, Tscharntke T (2016) Perceptions of cultural ecosystem services from urban green. Ecosyst Serv 17:33–39

Mathey J, Rößler S, Lehmann I, Bräuer A, Goldberg V, Kurbjuhn C, Westbeld A (2011) Noch wärmer, noch trockener? Stadtnatur und Freiraumstrukturen im Klimawandel. In: Bundesamt für Naturschutz (BfN) (ed) Naturschutz und Biologische Vielfalt, Issue 111:220 pp

Mathey J, Rößler S, Banse J, Lehmann I, Bräuer A (2015) Brownfields as an element of green infrastructure for implementing ecosystem services into urban areas. J Urban Plann Dev 141 (3):A4015001-1–A4015001-13

McDonald RI, Green P, Balk D, Fekete BM, Revenga C, Todd M, Montgomery M (2011) Urban growth, climate change, and freshwater availability. PNAS 108(15):6312–6317

McPherson EG (1997) Atmospheric carbon dioxide reduction by Sacramento's urban forest. J Arboric 24(4):215–223

Melichar J, Vojáček O, Rieger P, Jedlička K (2009) Measuring the value of urban forest using the Hedonic price approach. Reg Stud 2:13–20

MEA—Millennium Ecosystem Assessment (2005) Ecosystem and human well-being: scenarios, vol 2. Island Press, Washington DC

Müller N, Abendroth S (2007) Empfehlungen für die Nationale Strategie zur Biologischen Vielfalt in Deutschland. Naturschutz und Landschaftsplanung 39:114–118

Newton J (2007) Well-being and the natural environment: a brief overview of the evidence. United Kingdom Department of Environment, Food and Rural Affairs Discussion Paper 20 Aug 2007

Nowak DJ (2006) Air pollution removal by urban trees and shrubs in the United States. Urban Forest Urban Greening 4:115–123

Nowak DJ, Hoehn RE III, Crane DE, Stevens JC, Leblanc-Fisher C (2010) Assessing urban forest effects and values, Chicago's urban forest. Resource Bulletin NRS-37. Department of Agriculture, Forest Service, Northern Research Station, Newtown Square

Oke TR (1973) City size and the urban heat island. Atmos Environ 7:769–779

Panduro TE, Veie KL (2013) Classification and valuation of urban green spaces—a hedonic house price valuation. Landscape Urban Plann 120:119–128

Patz JA, Campbell-Lendrum D, Holloway T, Foley JA (2005) Impact of regional climate change on human health. Nature 438:310–317

Paul MJ, Meyer JL (2008) Streams in Urban Landscape. In: Marzluff JM, Shulenberger E, Endlicher W, Alberti M, Bradley G, Ryan C, ZumBrunnen C, Simon U (eds) Urban ecology. An international perspective on the interaction between humans and nature. Springer, New York, pp 207–231

Pauleit S (2010) Kompakt und grün: die ideale Stadt im Klimawandel. Grünflächen gegen Hitze in der Stadt. Themenheft Stadtklima. Garten + Landschaft 4:12–15

Rauterberg H (2013) Essbare Stadt: Lasst es euch schmecken! DIE ZEIT 36/2013. www.zeit.de/2013/36/urban-gardening-essbare-stadt. Accessed 9 Oct 2015

REGKLAM-Konsortium (ed) (2013) Integriertes Regionales Klimaanpassungsprogramm für die Region Dresden. Grundlagen, Ziele und Maßnahmen. REGKLAM-Publikationsreihe, Issue 7. Rhombos, Berlin

Ren QW, Wang C, Qie GF et al (2006) Airborne particulates in urban greenland and its relationship with airborne microbes. Urban Environ Urban Ecol 19(5):22–25

Richardson EA, Pearce J, Mitchell R, Kingham S (2013) Role of physical activity in the relationship between urban green space and health. Public Health 127(4):318–324

Rook G (2013) Regulation of the immune system by biodiversity from the natural environment: an ecosystem service essential to health. Proc Natl Acad Sci U S A 110:18360–18367

Rowe DB (2011) Green roofs as a means of pollution abatement. Environ Pollut 159:2100–2110

Rütten A (2001) Self reported physical activity, public health and perceived environment: results from a comparative European study. Epidemiol Community Health 55:139–146

Sæbø A, Popek R, Nawrot B et al (2012) Plant species differences in particulate matter accumulation on leaf surfaces. Sci Total Environ 427:347–354

Santos JML (1998) The economic valuation of landscape change: theory and policies for landscape conservation. Edward Elgar, Cheltenham

Schär C, Jendritzky G (2004) Climate change: hot news from summer 2003. Nature 432:559–560

Schär C, Vidale PL, Luthi D, Frei C, Haberli C, Liniger MA, Appenzeller C (2004) The role of increasing temperature variability in European summer heatwaves. Nature 427:332–336

Scherber K, Langner M, Endlicher W (2013) Spatial analysis of hospital admissions for respiratory diseases during summer months in Berlin taking bioclimatic and socio-economic aspects into account. Erde 144(3):217–237

Schmidt C, Seidel M, Großkopf F (2014) Entwicklung einer Methodik für die Ermittlung stadtspezifischer Richtwerte für die quantitative und qualitative Ausstattung mit öffentlich nutzbarem Grün in Dresden. TU Dresden, Dresden

Schröter F (2015) Orientierungswerte, Homepage Dr. Schröter, as of 13 Apr 2015. http://www.dr-frank-schroeter.de/planungsrichtwerte.htm. Accessed 1 Nov 2016

Semenza JC, Rubin CH, Falter KH, Selanikio JD, Flanders WD, Howe HL, Wilhelm JL (1996) Heat-related deaths during the July 1995 heat wave in Chicago. N Engl J Med 335:84–90

Shashua-Bar L, Hoffman ME (2000) Vegetation as a climatic component in the design of an urban street: an empirical model for predicting the cooling effect of urban green areas with trees. Energy Build 31:221–235

Shi M (1998) From imperial gardens to public parks: the transformation of urban space in early 20th-century Beijing. Modern China 24(3):219–254

Smyth R, Mishra V, Qian X (2008) The environment and well-being in urban China. Ecol Econ 68 (1):547–555

Spangenberg J (2014) China in the anthropocene: culprit, victim or last best hope for a global ecological civilisation? BioRisk 9:1–37

Spash CL (2000) The Concerted Action on Environmental Valuation in Europe (EVE): an introduction. Environmental Valuation in Europe (EVE), Cambridge Research for the Environment, Cambridge

Speak A, Tothwell J, Lindley S et al (2012) Urban particulate pollution reduction by four species of green roof vegetation in a UK city. Atmos Environ 61:283–293

Stanners D, Bourdeau P (1995) Europe's environment: the Dobris assessment. European Environment Agency, Copenhagen, pp 261–296

Stigsdotter UK, Ekholm O, Schipperijn J, Toftager M, Kamper-Jørgensen F, Randrup TB (2010) Health promoting outdoor environments: associations between green space and health, health-related quality of life, and stress based on a Danish National Representative Survey. Scand J Public Health 38(4):411–417

Streutker DR (2003) Satellite-measured growth of the urban heat island of Houston, Texas. Remote Sens Environ 85:282–289

Stülpnagel AV (1987) Klimatische Veränderungen in Ballungsgebieten unter besonderer Berücksichtigung der Ausgleichswirkung von Grünflächen, dargestellt am Beispiel von Berlin (West). Dissertation FB 12 (Landschaftsentwicklung), TU Berlin, Berlin 173 pp

Sturm R, Cohen D (2014) Proximity to urban parks and mental health. J Ment Health Policy Econ 17(1):19–24
Swanwick C, Dunnett N, Woolley H (2003) The nature, role, and value of green space in towns and cities: an overview. Built Environ 29(2):94–106
Taha H, Akbari H, Rosenfeld A (1991) Heat island and oasis effects of vegetative canopies: micro-meteorological field-measurements. Theoret Appl Climatol 44(2):123–138
Takano T, Nakamura K, Watanabe M (2002) Urban residential environments and senior citizens' longevity in megacity areas: the importance of walkable green spaces. J Epidemiol Community Health 56(12):913–918
Tan J, Zheng Y, Tang X, Guo C, Li L, Song G, Zhen X, Yuan D, Kalkstein A, Li F, Chen H (2010) The urban heat island and its impact on heat waves and human health in Shanghai. Int J Biometeorol 54(1):75–84
Tchepel O, Dias D (2011) Quantification of health benefits of related reduction of atmospheric PM10 levels: implementation of population mobility approach. Int J Environ Health Res 21:189–200
TEEB—The Economics of Ecosystems and Biodiversity (2010) Chapter 5: The economics of valuing ecosystem services and biodiversity
TEEB—The Economics of Ecosystems and Biodiversity (2011) TEEB manual for cities: Ecosystem services in urban management
TEEB DE—Naturkapital Deutschland (2016) Ökosystemleistungen in der Stadt – Gesundheit schützen und Lebensqualität erhöhen. In: Kowarik I, Bartz R, Brenck M. (eds) Technische Universität Berlin; Helmholtz-Zentrum für Umweltforschung – UFZ. Berlin, Leipzig
Tong C, Feagin RA, Lu J, Zhang X, Zhu X, Wang W, He W (2007) Ecosystem service values and restoration in the urban Sanyang wetland of Wenzhou, China. Ecol Eng 29:249–258
Tong MK, Gao JX, Tian MR et al (2015) Subduction of PM2.5 by road green space in Beijing and its health benefit evaluation. China Environ Sci 35(9):2861–2867
TURENSCAPE (2011) Qunli stormwater park: a green sponge for a water-resilient city. http://www.turenscape.com/en/project/detail/435.html. Accessed 25 Aug 2011
Tyagi V, Kumar K, Jain VK (2013) Road traffic noise attenuation by vegetation belts at some sites in the Tarai region of India. Arch Acoust 38(3):389–395
Ulrich RS, Simons RF, Losito BD, Fiorito E, Miles MA, Zelson M (1991) Stress recovery during exposure to natural and urban environments. J Environ Psychol 11(3):201–230
UNEP—United Nations Environment Programme (2010) UNEP Environmental assessment Expo 2010 Shanghai, China. http://www.unep.org/pdf/SHANGHAI_REPORT_FullReport.pdf. Accessed 3 Jan 2017
UNPD—United Nations Procurement Division (2012) World urbanization prospects: the 2011 revision, New York
Upmanis H, Eliasson I, Lindqvist (1998) The influence of green areas on nocturnal temperatures in a high latitude city (Göteborg, Sweden). Int J Climatol 18:681–700
URGE-Team (2004) Making greener cities—a practical guide. UFZ-Bericht 8 (Stadtökologische Forschungen 37), UFZ Leipzig-Halle GmbH, Leipzig. http://www.urge-project.ufz.de/index.html
US EPA—United States Environmental Protection Agency (2004) Incorporating emerging and measures in a State Implementation Plan (SIP). US Environmental Protection Agency, Research Triangle Park, NC. http://www.epa.gov/ttn/oarpg/t1/memoranda/evm_ievm_g.pdf
Van Praag B, Baarsma B (2005) Using happiness surveys to value intangibles: the case of airport noise. Econ J 115:224–246
Vanslembrouck I, Van Huylenbreock G, Van Meensel J (2005) Impact of agriculture on rural tourism: a hedonic pricing approach. J Agric Econ 56(1):17–30
Von Döhren P, Haase D (2015) Ecosystem disservices research: a review of the state of the art with a focus on cities. Ecol Ind 52:490–497
Voogt JA (2002) Urban heat island. In: Douglas I (ed) Causes and consequences of global environmental change. Wiley, Chichester, pp. 660–666

Vos PEJ, Maiheu B, Vankerkom J, Janssen S (2012) Improving local air quality in cities: to tree or not to tree. Environ Pollut 183:113–122
Wang L, Hasi E, Liu LY et al (2006) Effects of weather condition in spring on particulates density on conifers leaves in Beijing. Chin J Ecol 25(8):998–1002
Wang G, Jiang G, Zhou Y et al (2007) Biodiversity conservation in a fast-growing metropolitan area in China: a case study of plant diversity in Beijing. Biodivers Conserv 16:4025. doi:10.1007/s10531-007-9205-3
Wang X-A (2013) The impacts of 2003 extreme temperatures on mortality in Shanghai. East China Normal University, China
Ware JE, Kosinski M, Gandek B (2000) SF-36 health survey: manual and interpretation guide. QualityMetric Incorporated, Lincoln, RI, 1993
Weller B, Naumann T, Jakubetz S (eds) (2012) Gebäude unter den Einwirkungen des Klimawandels. REGKLAM-Publikationsreihe, Heft 3. Rhombos, Berlin
Welsch H (2006) Environment and happiness: valuation of air pollution using life satisfaction data. Ecol Econ 58:801–813
Whitfield J (2009) Seeds of an edible city architecture. Nature 459:914–915
WHO—World Health Organisation (2010) Urban planning, environment and health: from evidence to policy action. p. 119
WHO—World Health Organisation (2013) Review of evidence on health aspects of air pollution —REVIHAAP project: final technical report. Kopenhagen
Wolch JR, Byrne J, Newell JP (2014) Urban green space, public health, and environmental justice: the challenge of making cities just green enough. Landscape Urban Plann 125:234–244. doi:10.1016/j.landurbplan.2014.01.017
Wu S, Hou Y, Yuan G (2010) Valuation of forest ecosystem goods and services and forest natural capital of the Beijing municipality, China. Unasylva 234/235(61):28–36
Wu YX, Kang WX, Guo QH et al (2009) Functional value of absorption and purgation to atmospheric pollutants of urban forest in Guangzhou. Sci Silvae Sinicae 45(5):42–48
Wüstemann H, Kalisch D, Kolbe J (2016) Towards a national indicator for urban green space provision and environmental inequalities in Germany: Method and findings. SFB 649 Discussion Paper 2016-022
Wüstemann H, Kolbe J, Krekel C (2017) Health effects of urban green spaces: an empirical analysis. Natur und Landschaft 92(1):31–37
Wüstemann H, Kolbe J (2017a) Der Einfluss städtischer Grünflächen auf die Immobilienpreise: Eine hedonische Analyse für die Stadt Berlin. Raumforschung und Raumordnung 3:1–10
Wüstemann H, Kolbe J (2017b) Die Bewertung der ökonomischen Bedeutung urbaner Grünflächen mittels der Immobilienwertmethode. Geographische Rundschau 5:32–37
Xioa N (2015) Biodiversity and ecosystem services in Beijing City. Presentation at the 8th Sino-German workshop on biodiversity and conservation, Berlin, 18–19 June 2015
Xu Z, Wu J (2015) Residential greening environment and the physical and mental health of residents: the mediating role of life satisfaction. Psychology: techniques and Applications 6 (22):7–14 (in Chinese)
Yang W, Hyndman DW, Winkler JA, Viña A, Deines J, Lupi F, Luo L, Li Y, Basso B, Zheng C, Ma D, Li S, Liu X, Zheng H, Cao G, Meng Q, Ouyang Z, Liu J (2016) Urban water sustainability: framework and application. Ecol Soc 21(4):4
Yang J, McBride J, Zhou JX et al (2005) The urban forest in Beijing and its role in air pollution reduction. Urban Forest Urban Greening 3:65–78
Yang J, Yu Q, Gong P (2008) Quantifying air pollution removal by green roofs in Chicago. Atmos Environ 42:7266–7273
Yang J, Zhou J (2009) Comparison of vegetation cover in Beijing and Shanghai. A remote sensing approach. In: McDonnell MJ, Breuste JH (eds) Comparative ecology of cities and towns. Cambridge University Press, UK
Yin H, Xu J (2009) Spatial accessibility and equity of parks in Shanghai. Urban Stud 6:71–76
Zhang X, Chen J, Tan M, Sun Y (2007) Assessing the impact of urban sprawl on soil resources of Nanjing city using satellite images and digital soil databases. Catena 69:16–30

Zhang B, Xie GD, Zhang CQ et al (2012a) The economic benefits of rainwater-runoff reduction by urban green spaces: a case study in Beijing, China. J Environ Manage 100:65–71

Zhang B, Gaodi X, Bin X, Canqiang Z (2012b) The effects of public green spaces on residential property value in Beijing. J Resour Ecol 3(3):243–252

Zhang B, Xie GD, Gao JX et al (2014) The cooling effect of urban green spaces as a contribution to energy-saving and emission-reduction: a case study in Beijing, China. Build Environ 76:37–43

Zhao ST, Li XY, Li YM (2014) The characteristics of deposition of airborne particulate matters with different size on certain plants. Ecol Environ Sci 23(2):271–276

Zhou SZ (1990) Five island effects of shanghai urban climate. Science in China Series B Chemistry Life Sciences & Earth Sciences 33(1):67–78

Zhou N, He G, Williams C, China Energy Group (2012) China's Development of Low-Carbon Eco-Cities and Associated Indicator Systems. Ernest Orlando Lawrence, Berkeley National Laboratory, LBNL-5873E

Zhou S, Wang X (1996) The urban dry island and urban moisture island effects of Shanghai atmospheric environment. J East China Normal Univ Nat Sci 68–80

Zhou W, Qian Y, Li X, Li W, Han L (2013) Relationships between land cover and the surface urban heat island: seasonal variability and effects of spatial and thematic resolution of land cover data on predicting land surface temperatures. Landscape Ecol 29:153–167

Zippel S (2016) Urban Parks in Shanghai. Study of Visitors' Demands and Present Supply of Recreational Services. Dresden, TU Dresden, Faculty of Environmental Sciences, Master Thesis

Chapter 4
Options and Challenges for Implementing Green Spaces in Urban Development

Ralf-Uwe Syrbe and Jiang Chang

Providing highly qualitative, healthy green areas in densely populated cities is often a serious challenge. Cities need to search for new and innovative solutions to increase public space with vital vegetation that we call urban green space. Not only parks, lawns and roadside green belong to this category, but also rivers, creeks, ponds, and lakes together with their riparian biotopes are important parts of it. And there are many more opportunities to introduce vegetation into housing areas (e.g., green roofs, green facades) in order to enhance human health and quality of life. This chapter gives an overview of the possibilities and instruments and will go into depth on some particularly interesting issues, using selected examples in Germany and China at the end.

4.1 Instruments of Green Space Development

This section briefly explains the main levels and issues of city planning in China (Sect. 4.1.1) and Germany (Sect. 4.1.2) as far as they are important for green space development. Formal and informal planning instruments are presented here. Additionally, Sect. 4.1.3 comprises economic instruments in both countries, together with a box that outlines the split wastewater fee as a usual instrument of soil sealing mitigation in Germany as an example.

R.-U. Syrbe (✉)
Leibniz Institute of Ecological Urban and Regional Development, Dresden, Germany
e-mail: r.syrbe@ioer.de

J. Chang
China University of Mining and Technology, Xuzhou, China
e-mail: changjiang102@163.com

K. Grunewald et al. (eds.), *Towards Green Cities*, Cities and Nature,
DOI 10.1007/978-3-319-58223-8_4

4.1.1 *Urban Planning System in China with Regard to Green Space*

Jiang Chang, Tinghao Hu and Pingjia Luo

In China, a systematic green space development and planning system was established in the early twenty-first century and now forms a complete top-down regulation framework. From different planning domains, five legally defined types of plans at different scales all contribute to fulfilling green space development and protection aims: The Plan for National Economic and Social Development, the environmental protection and ecological control programs at the national level, the Land-Use Plan, Urban Master Plan, and the Urban Green System Plan. In addition, non-statutory plans and urban gardening movements are also important methods and incentives toward green space development and green city construction. Instruments of the green space development and planning system in China are shown in Fig. 4.1.

Plan for National Economic and Social Development

The Plan for National Economic and Social Development, also called the 'Five-Year Plan', condenses the specific arrangements of all issues of national economic and social development. Since 1953, China has issued 13 Five-Year Plans, which covered economic and social development, industry, IT, ecological and environmental protection, etc. On the aspect of improving green space development, this plan mainly acts as a steering wheel and determines the aggregate indicators at the national scale.

Land-Use Plan

The Land-Use Plan, a programmatic document of urban and rural construction and land management, is among all mentioned plans one of the strictest land management means. In accordance with the administrative region, the Land-Use Plan

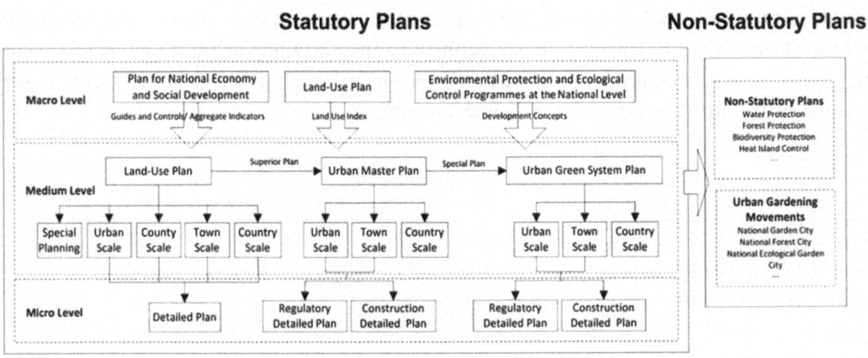

Fig. 4.1 Instruments of the green space development and planning system in China. © *Draft* Thinghao Hu

can pertain to the national, provincial, urban, county, and country levels. Its planning objects are all land resources and land-use types.

The Land-Use Plan and Urban Green System Plan overlap in content. While the names of green space types are different in the two plans, the Land-Use Plan covers all types of urban green space. It not only covers 'recreational green land' in the 'construction land' category, but also green space in other land-use types. As this plan is superior to the Urban Green System Plan, in addition to clearly defining urban green space, the Land-Use Plan plays more of a guiding and controlling role for unutilized land to be integrated as an important part of the urban green system.

The Land-Use Plan plays a definite role in planning urban green space, which consists in controlling the quantity, function, and layout of farm land, construction land, and unutilized land. As a result, it directly determines the scale, function, and structure of green space in urban areas.

Urban Master Plan

The Urban Master Plan is the framework plan for urban development. It is the comprehensive planning of urban economic and social development, land use, spatial layout, as well as urban management. Based on the land-use index determined by the Land-Use Plan, the Urban Master Plan will further determine the layout and form of green space in urban areas. This plan contributes to improving urban green space development by controlling and managing the greening in construction area. The Urban Master Plan has a complete set of planning programs from the urban scale to the country scale.

Urban Green System Plan

The Urban Green System Plan is one of the important special plans of the Urban Master Plan. In 2002, the Ministry of Construction (today called Ministry of Housing and Urban–Rural Development) issued the *Urban Green Space Classification Standard* and the *Urban Green Space System Plan Outline (trial).* Then, the Urban Green System Plan was created as a special independent (legally binding) part of the Urban Master Plan. On the basis of the urban character, the development goals and the land-use layout set by the Urban Master Plan, the Urban Green System Plan is designed to formulate target values of urban green space and defines types of landscape and green system at different scales. As a special plan of the Urban Master Plan, the Urban Green System Plan further implements subindicators and detailed planning requirements of the Urban Master Plan. It directly guides and controls spatial function, structure, form, and tree species of types of green spaces. Within the scope of the urban planning area, the planning targets of the Urban Green System Plan include: public parks, productive plantation areas, green buffers, attached green spaces, and other green spaces (Table 4.1).

Table 4.1 Green space classification of China

Category code	Green space types	Definition	Subordinated types
G1	Public park	Open to the public mainly with leisure function and with other functions of ecology, beautification and disaster prevention, etc.	Community park, theme park, linear park, roadside green space
G2	Productive plantation area	Provides nurseries, flower beds, humus, etc., for urban greening	–
G3	Green buffer	Green space with the function of health, isolation, and security, including greenbelt, grid and pipeline corridor, windbreaks, protection forest	–
G4	Attached green space	Green space attached to other categories of construction land type	Green space in residential land, commercial and public facilities land, industrial land, warehouse land, transportation land, road, street and square land, municipal utilities land, specially designated land
G5	Other green space	Directly affects urban environment quality, human well-being, urban landscape and biodiversity protection	Scenic spot, water conservation district, country park, forest park, natural conservation, scenic forest, urban greenbelt, animal park, wetland

Source: Ministry of Construction (2002)

Non-statutory Plans and Urban Gardening Movements

At the urban scale, government departments will direct research and development (R&D) institutions as well as colleges and universities to compile some non-statutory plans closely related to the development and protection of urban green space, such as a Key Ecological Function Areas Plan, a Biodiversity Conservation Plan, a Wind Corridor Plan, etc. (see Sect. 4.3.6). These plans are necessary supplements of the Urban Master Plan and the Urban Green System Plan. Meanwhile, the law and its enforcement remain to be strengthened.

In addition, a series of urban gardening campaigns have been rapidly carried out in China, like National Garden City, National Forest City and National Ecological Garden City. The government departments set indicators for different types of model cities, provide subsidies and support policy innovations if cities meet the standards and are elected as the corresponding model city. Urban gardening campaigns not only help to protect and develop urban green space and the environment, but also to promote awareness and appreciation of the city, as well as awareness of responsibility and of environmental protection among the citizens.

4.1.2 *Urban and Green Space Planning in Germany*

Stefanie Rößler and Alice Schröder

In addition to strategies, in terms of overall objectives and approaches (see Sect. 2.2), different formal and informal instruments of spatial planning and also sectoral planning can be used to plan and to implement requirements, objectives, and concrete measures of green space development in urban realities. In this chapter, first, the German system of planning and the corresponding planning instruments will be introduced. Second, recommendations will be formulated how individual measures can be implemented at the appropriate planning scales.

In Germany, an overall system of spatial planning exists to address spatial development at different scales. Based on sectoral legal frameworks, particularly nature protection and environmental directives, a number of appropriate instruments are available to address environmental issues at the various planning scales (Fig. 4.2). In particular, the formal instruments of spatial planning at the state, regional, municipal, and neighbourhood scale are supplemented by instruments of nature protection (landscape planning). To deal with existing urban areas and the building stock, formal instruments of urban redevelopment are available. Additionally, further instruments of environmental planning can be applied to promote urban green spaces. Besides the legally binding instruments, a variety of informal instruments at the regional and urban scale can be applied by cities individually (see below)

Land-Use Planning and Urban Redevelopment

In Germany, the objectives, purposes, and instruments of spatial planning are regulated in the **Spatial Planning Law** (federal state and regional level) and the **Federal Building Code** (local level). The overall aim of spatial planning is to coordinate requirements of land use by ensuring sustainable development and equal living conditions in the entire country. The task of the land-use planning in urban areas, based on the Federal Building Code, is to prepare and to guide the land use of municipalities following the principles of sustainable development. In particular, the environment, the natural features and challenges of climate change, but also the urban design quality should be addressed. The protection of existing and the development of new green spaces is an important part of this (Heiland et al. 2016; Rößler and Albrecht 2015; Wende et al. 2010).

Regional Plan

For regions as parts of federal states in Germany, **regional plans** are obligatory to coordinate the land-use demands resulting from agriculture, forestry, settlement, infrastructure, but also nature protection as well as greenbelts and green breaks. In terms of developing urban–regional green space systems, of guiding settlement activities to protect natural areas, and of enabling habitat networks, regional plans are important instruments to coordinate demands of neighboring municipalities.

Zoning Plan and Local Construction Plan

At the municipal level, two instruments are available to coordinate land-use decisions and future urban development. The **zoning plan/land-use plan** comprises the entire area of the municipality, including the settlement, but also open space areas such as agricultural land, forests and green spaces. Thus, the overall structure of the green space system, the distribution of different types of green spaces and their legal status are defined by zoning plans. The regulations of the zoning plan are obligatory for the municipal administration. Concrete planning decisions have to be harmonized with this plan. At the district/site level, **local construction plans** are created to regulate concrete building activities relating to density, building structure, green space ratio and functions (see Box 4.6).

The professional basis for green space development at both planning scales is provided by the corresponding obligatory instruments of landscape planning (see section "Landscape planning"). The requirements of nature conservation and landscape management have to be included in the weighting process that is obligatory in zoning plans and local construction plans. This means that the suggested aims and measures have to be taken in account equally. If they are not considered, appropriate reasons have to be provided. Objectives of landscape planning instruments become binding if they are integrated into the plans of spatial planning.

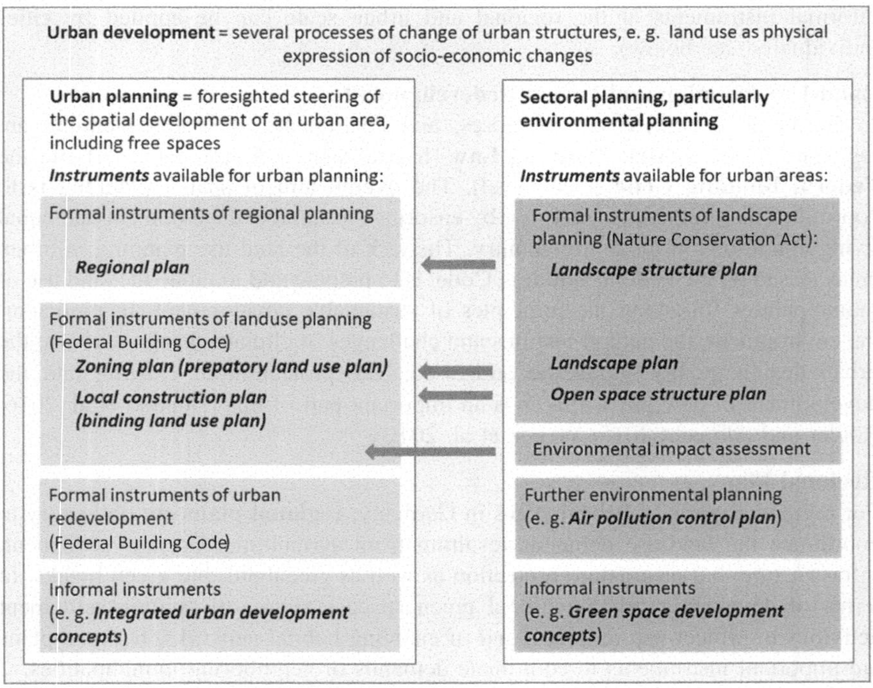

Fig. 4.2 Overview of planning instruments in Germany. © Stefanie Rößler

Therefore, it is essential to provide sound and high-quality municipal landscape plans.

In local construction plans, the following categories of green spaces can be determined:

- Public and private green spaces (parks, allotment gardens, sports grounds and playgrounds, cemeteries)
- Water areas
- Areas for forestry and agriculture
- Areas for protection, management and development of nature.

To create private green spaces near residential buildings part of the building activities. They are regularly not included in the plan, but obligatory due to benchmarks for building structures. In these plans, some basic requirements regarding the design, plant selection, and sealing ratio can be defined. Concrete design planning has to be carried out in the construction and building process, under the responsibility of the investor.

Redevelopment of Existing Urban Areas

As urban development in Germany mainly takes place in existing neighborhoods and settlement structures, urban renewal, in terms of redevelopment of existing urban areas, plays a big role in everyday urban planning. Due to the (also quantitative) relevance of the existing building stock, green space development and the provision of ecosystem services need to be addressed within these renewal strategies (BfN 2015).

The regulatory framework for dealing with existing urban areas is formulated in the Federal Building Code. There it is stipulated that urban renewal and retrofitting serve the well-being of the urban population. In particular, the urban and building structure should serve the general social, health, economic, and cultural requirements of urban life and also fulfill demands of environmental protection. By considering ecosystem services provided by green spaces, these requirements can be addressed. Thus, green space development is a crucial element of urban renewal (Heiland et al. 2016).

Different instruments for urban renewal are available by law, tailored for the differing urban renewal challenges (retrofitting of existing neighbourhoods, transformation of neighbourhoods, social challenges). Additionally, state funding programs are available to promote these renewal and redevelopment efforts. Recently, promoting biodiversity and green spaces has become one priority of these programs. Investments in public green spaces are worthwhile. Following public initiatives in distressed urban neighbourhoods, private investments in the housing stocks are encouraged (Fig. 4.3).

Scale	Urban-regional scale	Municipal/City scale	District/Neighbourhood/Site scale
Issues	Guiding of new settlement Green space network Fresh air corridors Habitat network Recreation	Green space network Distribution of green spaces Accessibility of green spaces Green space provision Habitat diversity Ecosystem services provision	Function and design of green spaces Ecosystem services provision Species diversity Multifunctionality Maintenance
Instruments	Formal: Regional plan Landscape structure plan Informal: Regional park concepts	Formal: Zoning plan Landscape plan Informal: Municipal biodiversity strategies Integrated urban development concepts Green space development strategies	Formal: Local construction plan Open space structure plan Informal: Instruments of urban renewal/ re-development

Fig. 4.3 Issues of green space planning and appropriate planning scales/instruments. © Stefanie Rößler

Landscape Planning
Landscape planning is the central and well-established planning instrument of prevention-oriented nature conservation and landscape management in Germany. It has been an essential part of the **German Act on Nature Conservation and Landscape Management** since 1976.

The primary **objectives** of landscape planning are the specification of the purposes of nature conservation and landscape management at the local and supralocal level and the identification of applicable requirements and measures for achieving those purposes. Therefore, landscape plans contain information on:

- the existing and anticipated status of nature and landscape,
- the specified objectives of nature conservation and landscape management,
- an assessment of the existing and anticipated status of nature and landscape on the basis of these objectives, including any possible conflicts,
- requirements/aims and measures to implement the specified objectives of nature conservation and landscape management,
- monitoring.

The **tasks** of landscape planning include the provision of information and standards for the implementation of biotope networks, sustainable land use and urban development, for the protection of biodiversity and for safeguarding landscape characteristics, etc. (BfN 2008). It also supports the assessment of the environmental impact of programs, plans, and measures of other disciplines, the application of compensation mechanisms and the enforcement of European nature conservation and water legislation. Landscape plans provide information for nature conservation authorities, planning, approval, and licensing authorities as well as for various organizations, land users, and citizens. The contents of landscape planning have to be considered in all planning and administration processes that affect nature and landscape. Thus, landscape planning supports fast and reliable decision-making on the basis of specified information on the condition of nature and landscape as well as objectives for further development.

In parallel to spatial planning, landscape planning addresses different spatial levels with different scales, information, and detail. There is no landscape plan at the national level. At the federal level, **landscape programs** have been drawn up for almost all federal states (so-called Bundesländer). At the regional level, there exist regional **landscape structure plans** for virtually all parts of Germany. At the municipal level, local **landscape plans** currently exist for almost half the area of Germany and are being prepared for approximately one-fifth of the area of Germany (BfN 2012). At the level of project sites, **open space structure plans** specify the layout of the respective planning area. The legal obligations and binding character of landscape plans at different spatial levels vary across the federal states.

At the level of the city, the local **landscape plan** is an important tool for analyzing and planning the green and open spaces of the city. This is at least as important as the development of buildings and gray infrastructure. A good local landscape plan provides all important information on nature and the environment and helps to assess the environmental and spatial compatibility of plans and projects of urban development (Schröder et al. 2016). Furthermore, it sets objectives and measurements for the protection, maintenance, and development of nature and urban green spaces.

Landscape planning at the city level provides a spatially specified and politically legitimated concept for long-term urban development with special regard to natural balance, ecosystem services, and quality of life. Beyond traditional issues such as species, biotopes, water, soil, climate and recreation, it is important for local landscape plans to address current challenges such as the integration of green and gray urban development (so-called double urban development, Böhm et al. 2016), health, and provision and accessibility of urban green spaces.

Last but not least, it is important to understand that a local landscape plan is not a final product but rather a modular and cross-sectoral instrument which constantly reacts to current needs. Hence it should be understood as an ongoing planning process.

Environmental (Impact) Assessment

In accordance with the directives of the European Union to ensure that the environmental implications of decisions are taken into account before the decisions are made, public plans and programs also have to undergo a strategic environmental assessment (SEA). Thus, to address and meet the environmental requirements of spatial development, an environmental impact assessment is obligatory for regional and municipal land-use planning.

Informal Instruments

In addition to formal instruments, informal approaches can address specific challenges of urban development. Through these voluntary plans or programs municipalities are committing themselves to individual targets and strategies. Although not legally binding, these informal approaches have advantages: Specific and individual issues can be addressed without following the strict agendas and procedures of the legal instruments. Intersectoral, multi-scale, and multi-stakeholder cooperation can be enhanced. Through broad (public) discussions, a broad understanding and in the end support can be ensured.

In particular, cross-cutting green space issues can be the subject of such plans, as (in Germany) no formal instrument exists by which these issues can be addressed explicitly. Accordingly, some cities are developing dedicated green space development strategies (Rößler et al. 2016). Recently, tailored municipal biodiversity strategies have been developed to address the specific requirements of urban habitat and species biodiversity (Sect. 2.2.3). Also, green space development is recognized as an important part of (informal) integrated urban development concepts (Bläser et al. 2012).

4.1.3 Economic Instruments and Financing for Green Space in Cities

Bernd Hansjürgens, Hongxuan Zhou, Jiang Chang and Tinghao Hu

There is a variety of stakeholders that have different interests and ideas for the space available in urban areas. They all will opt for those alternatives that offer them the best value for money given the applicable rules and conditions of the urban area in which they operate (Droste et al., in print). Against this background, creating incentives with economic instruments may change these conditions, ultimately stimulating actors to pursue another project alternative that now seems best to them. For example, a separate charge on rainwater poured into sewage systems can offer incentives to avoid soil sealing and can thus create an incentive to sustain urban green areas (Rüger et al. 2015, cf. Box 4.1 and 3.3.2). The same is true for a wastewater fee that is oriented to the sealed natural ground (Geyler et al. 2014).

Basically, there are two classes of economic instruments which may help implement green spaces in urban areas.

Price-based economic instruments: Prices (in the form of fees, charges, taxes, or subsidies) incentivize a particular behavior through either increasing the costs of undesired environmental behavior, e.g., using green space for the sealing of soils and for housing developments, or by compensating for (additional) costs of implementing urban green spaces, such as through subsidies to install green roofs or to enhance existing green spaces to provide more (and other) ecosystem services. A possible application of such price-based instruments is through an incentive-oriented design of existing charges (e.g., municipal fees for water services) or levying new ones (e.g., levying urban water charges). Charges are meant to change the price of using an ecosystem service to reflect the full cost of its provision. Fees for water services are usually based on the principle of cost recovery. In the past, these charges often only included “technical” costs (i.e., investment, and operating and maintenance costs) incurred for the provision of water to users (Gawel 1995). However, there is leeway here, as the determination of underlying costs allows—at least in principle—to also include environmental and resource costs (Gawel 2016).

Box 4.1 Split waste water fee—a municipal fiscal instrument to promote ecosystem services in Germany (*by Martina Artmann*)

Soil sealing covers the land and its soil with artificial impervious and semi-impervious materials and has strong impacts on the supply of ecosystem services. Depending on the degree and material (e.g., concrete, grass pavers) of soil sealing, it disturbs the natural water cycle when it rains. Since soil sealing prevents natural precipitation runoff, rainwater is diverted into sewer systems to reduce the risk of flooding and high water. The soil, which is usually filtering, storing and supplying the groundwater, is reduced in its function as a natural ditch, and soil sealing threatens the supply of regulating (e.g., storm water runoff) and supplying ecosystems services (e.g., fresh water supply).

To create incentives for residents to reduce sealing at their buildings and to use, depending on the current use of land, semi-impervious materials, German municipalities implement a split waste water fee. Usually, the costs arising for waste water disposal are calculated based on fresh water consumption, neglecting costs related to rainwater disposal. The calculation of the costs for rainwater disposal is based on the discharge coefficient (Ψ). Thus, the runoff rate of rainwater depends on the surface texture (see Table 4.2). The proportions of the various surface types differ depending on the development type.

To calculate the amount for the split waste water fee more easily, the city of Munich, Germany, structures its territory based on similar built-up types and assigns areal discharge coefficients to these types (see Table 4.3). The split waste water fee is then calculated by multiplying the property size and the discharge coefficient of the respective built-up type. The result is then multiplied by the fee rate, which is currently 1.30 €/m^2. The fee has to be paid annually by the residents. If the effective area related to the discharge coefficient is at least 400 m^2 or 25% smaller than the area determined in the calculated fee, residents can request an individual calculation and reduction of the fee (Municipal Sewage Works of the City of Munich 2012). Thus, the split waste water fee is an example of a fiscal instrument to incentivize residents toward greening their properties and to increase ecosystem services (see also Sect. 3.3.2).

Table 4.2 Discharge coefficient according to DIN 1986-100:2002-039 (Wessolek et al. 2014)

Land use cover	Discharge coefficient (Ψ)
Impervious surface	
Roof > 3° slope	0.8–1.0
Roof < 3° slope	0.8
Gravel roof	0.5–0.7
Extensive green roof <10 cm thickness	0.5
Concrete surface	0.9
Semi-impervious surfaces	
Precast concrete stone	0.7
Surfaces with pavement, with joint portion >15%	0.6
Waterbound coverings	0.3–0.5
Plastic sheeting, artificial green with drainage	0.6
Lawn area	0.3
Permeable surfaces	
Parks and vegetation areas	0.3
Gravel, fine gravel	0.6
Driveways and carports with grass pavers	0.15

Table 4.3 Areal discharge coefficient in Munich, Germany (Municipal Sewage Works of the City of Munich 2012)

Built-up types	Areal discharge coefficient (Ψ)
Single-family houses and broken up row house buildings	0.35
Compacted row house buildings and linear developments	0.50
Dense building development in the periphery of the downtown	0.60
Dense building development in the downtown and highly sealed commercial areas	0.90

Intergovernmental fiscal transfers are a specific form of 'prices' among jurisdictions: While fees, charges, and taxes create incentives for private land users, the criteria according to which the levied tax income is distributed among different state jurisdictions creates incentives for decision-making of public authorities. Typically, fiscal transfers are assigned on the basis of comparing population numbers of communities (as a proxy for fiscal needs) and their own tax revenues, thereby stimulating communities to increase the number of inhabitants, e.g., by dedicating land for development to keep property prices low. If a portion of tax revenue were to be distributed according to ecological criteria, this might create incentives for providing green infrastructures and Nature-Based Solutions (NBS).

Quantity-based economic instruments: In contrast to price-based approaches, quantitative instruments directly limit activities that impact natural areas, e.g., by setting a cap on the maximum amount of green fields to be developed. Depending on the overall determined number of development titles (which make up the 'cap' of the system), these development titles will be auctioned or allocated for free among potential developers. By making development titles tradable, a cost-efficient allocation of development can be assured, as those landowners able to realize the highest net benefits from development will buy up titles and develop their land. However, if such a system is to allow for targeted protection of specific green infrastructures, it has to be accomplished by land-use zoning (Schröter-Schlaack 2013, Santos et al. 2015).

Economic Instruments for Green Space Development in China

In China, the laws concerning the environment and resources, as well as related policy documents, have clauses to protect the environment and natural resources by means of economic instruments. For example, they contain formulations such as "integrate the environment and natural resources into the national economic accounting system," "tax policy plays an important role in integrated decision processes," "the combination of the sectoral development policies and macroeconomic policies with the protection of the environment and resources," "the combination of economic means, regulation and control means," and "the duty of pollution control and protection by developers who cause pollution and damage" (Xie and Song 2014).

Ecological environment compensation is the main method of financial instruments with respect to green space development and protection in China. The practice of ecological environment compensation began in the early 1980s and aimed to impose costs on the ecological impact and destruction caused by mining activities (Zhang 2005). In 1996, China's State Council proposed to establish an economic compensation mechanism that compensated for the use of natural resources and the cost of environmental restoration. It was set up according to the rule "polluter pays, user compensates, developer protects and destroyer restores."

Compensation for the ecological benefits of forests is one of the earliest financial instruments that were widely implemented with a view to green space development in China. In 1998, an ecological benefit compensation fund was set up to support the planting, maintenance and management cost of shelter forest belts and forests with special uses. In 2000, the *Regulation on the Implementation of the Forestry Law of the People's Republic of China* was issued and stipulated that "the owners of shelter forest belts and forests with special uses have the right to obtain compensation for ecological benefits." In 2001, one billion yuan from the ecological compensation fund was allocated by the State Council for the development and protection of forest resources (Zhou and Zhao 2014).

At the urban scale, the construction of green space was implemented strictly according to the Land-Use Plan, the Urban Master Plan, and the Urban Green System Plan. In accordance with the relevant planning contents and specifications, local governments will release tenders for green space design projects. Public units and enterprises participate in a competitive bidding process and then implement the construction work. Local governments will finally check and accept the green space construction work.

In terms of green space maintenance, administrative departments such as the Landscape Garden Bureau and the Forestry Bureau are mainly responsible for the maintenance work. There are three modes of specific maintenance management (Li 2010):

- Public unit management; greening institutions that belong to local government are fully in charge of green space maintaining work; or expenditure is allotted by fiscal tranche.
- Enterprise management. Through a bidding process, green space maintenance work is contracted to a qualified professional enterprise, and local government plays a role of management, supervision, and assessment.
- Dual-track management. Public unit management and enterprise unit management will coexist at the same time.

In most cities of China, enterprise management has become the mainstream urban green space maintenance mode (see Fig. 4.4).

While land use is strictly under the control of the Land-Use Plan and the Urban Master Plan, rewards and subsidies for vertical greening have become important incentives to increase urban green space. Although China has not issued subsidies and incentive policies at the national level yet, many cities already regard vertical greening as an important method to increase green space in urban areas. Shanghai is

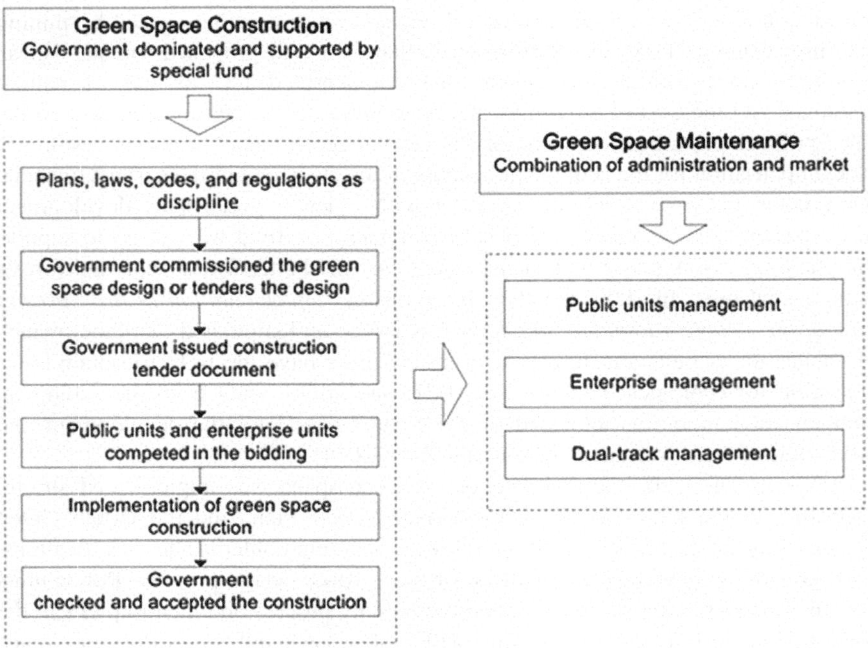

Fig. 4.4 Green space construction and maintenance system in China . © Thinghao Hu

the first city that introduced the skyrise greenery policy into local regulations (Box 4.5). In Shanghai, the subsidies for gardening roof greenery, assembled roof greenery and groundcover roof greenery are 200, 100, and 50 yuan/m^2, respectively. The maximum subsidy for a single greenery demonstration project can be up to 6 million yuan. In Beijing, the subsidy for skyrise greenery is 50–100 yuan/m^2 (Zhu et al. 2011). In Chengdu, medium high-rise buildings, multistory buildings and low-rise buildings are required to implement skyrise greenery. In addition, skyrise greenery is an independent part of the Urban Green System Plan, which has been integrated into the Urban Master Plan of Chengdu (Deng et al. 2010).

4.2 Options to Increase Green Spaces and Biodiversity in Cities

Ralf-Uwe Syrbe, Jiang Chang and Thinghao Hu

Implementing green infrastructure must be proactive to achieve success. Since nature needs time to develop and green space planning is often getting the short end of a deal, the relevant stakeholders have to be clear in advance about their objectives and designated implementing measures before they can actually act. To develop coherent green networks, the necessary sites should be reserved very early step by step. Long before a complete biotope network can become apparent, the foundation must be known and set by planning. Only in accordance with all stakeholders can urban green space be maintained and enhanced against the pressure of the real estate market, traffic development as well as the overwhelming housing and industrial land consumption. This section tries to provide a possible road map with a spectrum of suggestions and suitable starting points for developing green infrastructure in German and Chinese cities, which have been proven applicable and present appropriate examples as inspiration. The following Sect. 4.3 will deepen some selected recommendations that can only be shown here in brief in case studies from both countries.

German practitioners recommended the following steps for the improvement of urban green space in order to enhance biodiversity (Gärtner 2015): protection from housing and other development, reduction or renunciation of agrochemicals and turf, use of native plants and seeds, garden maintenance in harmony with nature, and transformation from lawn into meadows. The urban green infrastructure should be composed of several diverse biotopes and land-use types. It should unite private, common, and public areas but give access to all people and wherever possible also allow movement of wildlife. A diverse mosaic of green infrastructure elements in cities reflects the biodiversity of the surrounding rural landscape, enhanced by unaltered nature processes, but also reflecting the ideas of creative people.

The design of parks, gardens, and other types of urban green space has a long tradition in China. Classic Chinese gardens and scenic arrangements have been used for green space in Chinese cities for almost 2000 years. These gardens were based on the philosophical ideas of harmony and beauty according to Taoism, Confucianism, and Buddism (Ignatieva et al. 2015). Green spaces in China are well designed, and publicly accessible city parks are even a symbol of the new age of China's urbanity. At the beginning, these parks, called 'gongyuan', were the former royal gardens that have been slightly modified and opened for the public; but recently, several multi-functional new ones with modern landscape architecture have been designed in the growing Chinese cities (Küchler and Shao 2017). However, these areas are rarely near to nature or connected. Chinese plants, gardening designs, and constructions were adopted by European rulers already in the Middle Ages. Until today, the dense and artificial design of some Chinese parks is unique in the world. But also vice versa, European and Soviet design elements were introduced to China. Several uses that are common in Europe and America are rare in China by now, such as sports (sailing, swimming, diving) in urban water bodies. Likewise, the idea of nature and wilderness in cities is rather new in Asia with its long cultural traditions.

Altogether, the meaning of green space is quite different in China and Germany, so that both countries may learn from each other on several issues that are compiled below.

4.2.1 Alternative Types and Approaches for an Enrichment of Urban Green Space

Depending on the specific situation of a city, there are different ways to implement and optimize green space. Therefore, a considerable range of solutions exist. The classical types of green infrastructure such as parks, gardens, green yards, roadside green, urban grassland, arable land and forests, sports fields, and cemeteries belong to the standard repertoire of urban planning and green space design, which cannot be treated comprehensively within the scope of this book. But green space can also be near-natural remnants, religious sites, historical places, artificially designed facilities, recreation areas as well as naturally overgrown dumps, wastelands, disaster, or simply leftover areas. Thus, there are several solutions for enrichment of green space even in cities with a strong growth rate and spatial constraints. This section concentrates on alternative types and development approaches that may supplement the classical way of green space design. The main possibilities are given below; more detailed descriptions can be found in the case studies in Sect. 4.3.

Urban Biodiversity Hotspots—Nature Protection Core Areas
Mostly as remnants of pre-urban landscape, many cities contain original or near-natural biotopes. Mountains, woodlands, lakes, floodplains, and deep gorges may remain untouched by urban development and give a city a unique characteristic. These areas are the most valuable assets of nature and must be protected from development and transport projects. Some cities are proud of their valuable natural sites and glad not to have converted them in the past, such as Xuzhou of its mountain wood, or Leipzig of its floodplain forest (Fig. 4.5).

Fig. 4.5 Floodplain forest as a unique near-natural remnant within the major German city of Leipzig. © Ralf-Uwe Syrbe

Green Networks Connecting Urban Dwelling and Surrounding Landscape for Migration of Animals and Recreation of Inhabitants

People like to leave their vehicles in the garage and go—alone or with their family, perhaps also with a dog—into surrounding green space to recreate or exercise there. Therefore, not only a gray infrastructure consisting of streets and railways, but also green infrastructure with pathways is necessary. Such an alternative system of connected biotopes and trails (Fig. 4.13) provides added values: Wildlife can migrate, exchange of purified air is possible, and other functions are enhanced (see Chap. 3 and Sect. 4.2.2).

Green Connections Along Rivers and Creeks

Cities can restore formerly covered or neglected rivers in order to establish a new relationship with their watercourses using several innovative ways, as shown by Georgieva (2015). City planners, landscape designers, and society are asked to find ways to improve nature along the rivers, which should be opened and redesigned for nature and people. Green connections along rivers can serve as flood retention areas and the backbone of an ecological network. A common option is to rediscover abandoned industrial riverside buildings and restore so-called loft housing area there, to reintroduce the watercourses and thereby to increase the quality of life of the surrounding areas. But there are many more possibilities to turn riverside areas into new green spaces (Fig. 4.6).

Each solution to reintroduce nature along watercourses back into the heart of a city should be of high quality and fulfill the specific demands of residents. Among the most impressive examples is the Tanghe River (Box 4.2). Another example from Asia is the Kallang River—Bishan Park in Singapore, completed in 2012. The project to transform the Kallang river park emerged from the necessity to create more flood capacity for the river but also to enrich the functions of the green space. The vision is: Give the rivers space again. An industrial-style canal was removed to leave space for the naturally meandering river, providing access to the river during all of its level changes. With material from the canal's destruction, a hill was raised up as a viewpoint to the near-natural area and its urban surroundings. The result opened up the way that people can experience the water and interact among themselves. It connected the previously separated communities through the numerous recreation and sports activities within the park today.

Box 4.2: Tanghe River in Qinhuangdao City (Hebei Province, China; after Georgieva 2015)

Along the Tanghe River in Qinhuangdao, the 'Red Ribbon Park' has been designed. The river banks were saved from the fast-growing urban development with very little interventions. The park creation came just-in-time to handle the growing need for public space and the encroaching urban development. Before its transformation, it was taken over as a trash dump by the adjacent beach town and was unsafe due to the spread of shantytowns.

Major design elements are red ribbon runs through the park. They were designed to integrate lighting, seating and boardwalk, aiming at preserving the natural habitats while at the same time creating opportunities for recreational activities. Since its completion in 2008, the surroundings of the park have been urbanized and the population has higher demands for the maintenance of quality of the natural environment. The Red Ribbon Park responds to both the demands of nature and the public, providing a place for them to interact safely.

Fig. 4.6 Riverside greenery in the city center of Xuzhou. © Ralf-Uwe Syrbe

Green City Rings Surrounding the Intensely Settled Areas
Wherever a city district is built up too densely for implementing green spaces, it should be adjoined by a greenbelt providing a part of the missing greenery functions at least at the edge, and wherever possible with green fingers to connect the city center with its rural surrounding. Ring roads, abandoned industrial areas, and urban restructuring zones are starting points for creating such greenbelts. A well-developed forest belt surrounding a city can provide an important location factor, even enhancing the value of adjacent dwelling areas to a certain degree. A good example of implementing even several green rings around a dense city is the Shanghai case (Sect. 4.3.3).

Redesigning Brownfields as Green Spaces
Even though a city is prospering very well, there will appear brownfields/fallow land that cannot immediately be used again. The city administration should compile an overview of all fallow plots and their condition in order to continue using them and to offer the information for potential investors. A good brownfield management can not only reduce land take but can also create opportunities to enhance green spaces in the city. According to whether a brownfield can be reused very quickly or only later, an intermediate use for urban green is recommendable. Even without sophisticated upgrading, a fallow plot can be used for instance as a social meeting place, a community garden, a shrub area or coppice for bioenergy provision and climate regulation or at least as a temporarily undisturbed biotope (Mathey et al. 2015).

Green Upgrade of Post-mining Land and Brownfields
Mining is accompanied by environmental impacts as well as fundamental social reorganization, not only as long as resources are mined for decades but also when reserves are exhausted and the mines abandoned, which may have serious consequences for an entire city. The challenges with abandoned mine fields are similar in almost all mining regions (Wirth and Lintz 2007), but most crucial in post-mining cities.

However, the damaged post-mining land has important ecological potentials. Due to the long-term abandonment and little human disturbance, the ecosystem is in a state of self-recovery and free development, which provides survival space for threatened species and lays a foundation for the richness of species in these areas (Chang et al. 2011; Feng et al. 2016). Transforming post-mining land into agricultural and forest land can provide food, wood, and other natural resources for human beings. Established wildlife preserve and natural reserve areas have great positive effects on biodiversity. In the German Lusatia mining area, 15% of the former mining land was restored to natural reserve areas. This approach of ecological restoration is more common in European countries, while comparatively rare in China.

Post-mining land near urban areas is often used as a park, stadium, golf course, or fishing zone, which provide better cultural services. Examples are the concepts ‘Regional Park’ of the Ruhr Coal District (Siedlungsverband Ruhrkohlenbezirk,

SVR) from 1923 and 'Ruhr Open Space System' of the Ruhr Local Federation (Kommunalverband Ruhrgebiet, KVR) in 1985 of creating and maintaining a green space network in the German Ruhr agglomeration. In China, more and more parks being restored from coal mining subsidence areas have already become top choices for citizens to recreate and enjoy their leisure time (Box 4.3).

Box 4.3 Pan'an Wetland Park in Xuzhou, China (*by Jiang Chang, Tinghao Hu*)

Pan'an lake (Fig. 4.7) is located in Xuzhou, China, and is formed by vast stretches of coal mining subsidence areas with an area of 15.98 km^2. Since 2008, ecological restoration in coal mining subsidence areas has been one of the main tasks in urban construction in Xuzhou. In order to implement the restoration work in the Pan'an lake area, Xuzhou established a construction mode including basic farmland management, reclamation of subsidence areas, ecological restoration and landscape development in wetlands.

An afforestation and restoration project started in 2011. After 2 years of restoration, 4.3 km^2 of earthwork had been reclaimed and formed a continuous lake and water catchment areas. More than 0.8 km^2 of aquatic plants were planted in the west and northeast lake area. More than 100 kinds of trees, 200 varieties of shrubs, and 300 varieties of aquatic plants were planted. In addition to the ornamental tree species, there are a large number of crop plants, such as hawthorn, water chestnut, loquat, pomegranate, cherry, persimmon, etc. A vast wetland landscape was created that also meets tourist demand.

By the end of 2015, Pan'an Wetland Park received more than 3.6 million tourists. Its construction demonstrated how to transform a "burden" into an advantage. A series of large public activities, such as an international music festival, a long distance running race, film and television filming activities, have greatly enriched citizens' life.

Green Façades and Green Roofs

Green roofs and to a lesser extent green façades can reduce the need for air conditioning or heating by shading the surface and acting as an additional isolation layer in the building envelope. The aesthetic effects of green façades are well known and often employed, particularly in Southern Europe, as the Italian example in Fig. 4.8 shows impressively. By absorbing water from precipitation, green roofs smoothen the flood peak, preventing flood catastrophes and reducing the precipitation runoff in cities, and thereby the soil erosion of the non-built surfaces in cities. Green roofs help to lower the rain sewage charge in many German cities (Box 4.1).

Vegetation on roofs and façades serves to lower the pollutant level in cities through gas exchange and absorption. In particular, green shields created by vertical

Fig. 4.7 Pan'an wetland park and tourists. © Thinghao Hu

Fig. 4.8 Green façade in Sirmione, Italy. © Ralf-Uwe Syrbe

vegetation (ivy walls) take up CO_2, produce O_2, and collect particle matter deposition on their leaves. Some efforts and results of façade and roof greening are presented as an example from Shanghai (Box 4.5).

4.2.2 Enhancing the Functions of Green Infrastructure for the Benefit of People

Ralf-Uwe Syrbe, Jiang Chang, Thinghao Hu and Shanshan Feng

The most obvious urban green space type is the classical park, sometimes just as a small garden, sometimes extensive woods with lawns and benches along the pathways. The recreational value of urban green space depends not only on visual beauty but also on accessibility, usability, safety, silence, and air quality. Beyond the classical use as a picnic meadow, playground, walking arena and place for rest and contemplation, urban green space should be shaped in a way that it can fulfill several demands. By implementing multifunctional green infrastructure planning, multiple ecosystem services can be provided (Hansen and Pauleit 2014).

The green infrastructure is basically multifunctional. One site can provide several functions, which are sometimes competing but may also complement each other. The challenge is to create synergies of positive effects and to avoid trade-offs between ecosystem services. The importance of green space for policy will rise the more it can be connected with social structures, demands, and functionalities. Each part of green infrastructure has its own history and individuality, but all together may result in new functions and values for the system as a whole and thus for the quality of urban life. The following collection should give a rough idea of the several possibilities how green infrastructure can be constituted, what can be designed, and which different functions are available for an overall green city conception.

Recreation, Health and Sports

Facilities to enable active recreation can be included in parks and green spaces, even though they are small and unnatural. Offers to exercise in a green environment can be diverse: jogging trails, courses for inline skating and waveboarding are common in German cities; Chinese parks often host outdoor fitness facilities, which are well attended (see Fig. 4.9). Trails for biking and riding should be separated due to safety concerns when they are highly frequented. Particularly for the sporting youth, a climbing wall, obstacle courses and perhaps a half-pipe should not be missing in each city. Planners ought to think of winter activities in each suitable case. A sledding hill on a sloped meadow with a safe flat area at the end (without street or river) should be part of each hilly public green space if the climate allows it. For the warm seasons, fountains and bubblers provide agreeable chill. So-called "Kneipp gardens" are popular constructions in Germany that give people opportunities to get into contact with cold water that may strengthen the immune system and provide refreshment and cooling on hot summer days. Sebastian Kneipp (1821–1897) was one of the founders of naturopathic medicine in Germany and is well known for his special type of hydrotherapy, the application of water with different temperatures to several parts of the body, making use of the healing effects.

Fig. 4.9 Outdoor fitness arena in a small park in Shanghai city center. © Ralf-Uwe Syrbe

"Health is a state of complete physical, mental and social well-being and not merely the absence of disease or infirmity" (WHO 1946). The effects of green space on human health are diverse: Rittel et al. (2014) mentioned physical, social, psychological, aesthetic, and symbolic effects. But there can sometimes also be harmful impacts of urban green that should be avoided by the design and maintenance of green spaces as far as possible. Allergic diseases caused by pollen, particularly of birch, haze, and poplar are well known, but even more dangerous are ragweed (Ambrosia artemisiifolia) as a pollen allergen in Europe and giant hogweed (Heracleum giganteum) as a phototoxic risk plant, which is likely to invade poorly managed green spaces.

The implementation practice concentrates on the protection of human health against negative impacts (noise, air pollution) and refers to human health more indirectly by improvement of the beauty and recreational usability of urban green spaces. Altogether, it is important to secure daily quiet zones without traffic and industrial noise, but also night star arenas far from artificial light. The latter—nightly dark areas—are important for sensitive animals and people as well as for astronomy.

Green spaces are social meeting places in Europe and even more so in China. Well adopted are public barbecue areas and campfire places which can be used for free but which allay safety and environmental concerns. A well-conceived concept for such places and their municipal management (including cleanliness control and police care) can enhance the public green space and may increase the quality of life and the city's atmosphere considerably (Box 4.4).

Box 4.4 Dresden barbecue and campfire spaces near the Elbe river *by Ralf-Uwe Syrbe*

There are some places for barbeque and campfires in Dresden, defined by the city government. Lighting campfires and barbecuing in public green spaces is allowed exclusively in these places in order to protect nature and to ensure fire safety. An interactive version of the so-called thematic city map (see Fig. 4.10) shows these sites and explains the different permissions and restrictions. All the sites are paved and situated in an enjoyable environment (i.e., near the riverside). Barbecues are permitted at all these sites without preconditions, but a registration must be obtained before lighting a campfire. This rule is intended to prevent that several people want to light a fire simultaneously in the same place and one group is forced to shift outside the security zone for this reason.

Green for Elderly People

In China, urban parks have a special appeal for the elderly in the city (Hu and Shen 2014). For most of them, the biggest obstacle that aging brings is not inconvenience caused by physical degeneration, but psychological change and fuzzy self-awareness caused by retirement (Hu et al. 2016a, b). In China, the urban parks have become the elderly's most important places for leisure, recreation, sports, chatting, gathering, and various activities. They also provide the ideal platform for the realization of physical activities (Hu et al. 2016a, b). In Xuzhou, China, people aged 60 and above account for 1.6 million.

Yunlong Park, Xuzhou, was established in 1958, the first large-scale comprehensive park to be established after the foundation of the People's Republic of China. In 2007, the park was reconstructed. Considering the physical and mental characteristics of the elderly, the idea of 'design for access' was implemented in the reconstruction work. Three grades of roads were designed in the roads system. The first-grade road is fully barrier-free, so that wheelchairs and nonmotorized vehicles can directly get into the park. The first-grade road connects major scenic spots and three entrances located in the north, west, and south. The secondary and tertiary grade roads distribute to the various park attractions with high accessibility. Plenty of benches along the first-grade road provide enough places for rest. Warnings are painted in yellow on the stone steps to prevent falls. Lots of squares in the park provide the elderly with places to play Taiji, swords, and dancing. Yunlong Park has become an important meeting point for elderly people in Xuzhou (Fig. 4.11).

Green for Kids

Urban green space should be safe and useful for the entire family (Fig. 4.12). Besides the older people, special attention must be drawn to children and young people. Since their physical ability, wishes and requirements change very quickly, different equipment and creative offers are necessary to match these demands. Basic configurations are playgrounds featuring several preferably wooden play elements and a seating area for their parents, grandparents, or siblings. Very popular are

Fig. 4.10 Section of the so-called thematic city map of Dresden showing the allowed barbeque and campfire sites. Source: http://stadtplan.dresden.de/(S(5zgmykfw3hipyvximwjqoapg))/spdd.aspx?TH=UW_LAGERFEUER_GRILL

Fig. 4.11 In Xuzhou, Yunlong Park is used as a place for exercise, communications and activities, especially for the elderly. © Thinghao Hu

Fig. 4.12 Lakes and rivers are used as interesting places for children to experience and understand nature. © Thinghao Hu

trampolines and playing opportunities with water, sand, wooden barriers, and pumps. A hideout garden or labyrinth can encourage the joy of playing and imagination. Good ideas using 'nature-based solutions' are willow houses and ways that will sprout out and green themselves over time. So-called nature experience areas are to be recommended, where children can hide, play, and construct their own booths or dams (Pawlikowska-Piechotka 2011; Wolch et al. 2011). For instance, former garden areas with old fruit trees and overgrown lawns can be used in this way that requires a minimum of future maintenance and provides interesting possibilities for hiding, climbing, and constructing shacks.

Traffic in Green Space

A distinguished example of synergies between social needs, economy, and green infrastructure is the preparation of walking and cycling routes together with the completion of a green (biotope) network. Footpaths and cycleways through green spaces will not only provide enjoyment for people (Fig. 4.13). Shortcuts such as street underpasses, pedestrian bridges and passages through traffic-calmed areas would encourage people to prefer these ways as a car-free traffic alternative. While cycleways along the streets in our big towns are often dangerous, polluted and dull, a green alternative invites people to change. Asphalted main routes and accessible walks (for buggies, wheelchairs, wheeled walkers, cyclists, Segways) can be used even by disabled people and are safe in the cold season, too. For the latter, snow-plowing services may also be necessary. Today, eco-mobile rental stations (bicycles, Segways, handcarts) are common and well appreciated. For the future, solar charging stations should also be planned.

Fig. 4.13 Areas for species protection (old wood, stone cairn), footpaths and cycleways shorten the city's highways, encouraging people to move without a car in Dresden. © Ralf-Uwe Syrbe

Green Art and Culture

The urban green space can offer possibilities for artificial activities and experience. A stage in the green should not be missed in each great park, providing a place for concerts, dance events, and meetings of people. Wooden sculptures and artificial designed benches are favored elements, which often derive from competitions or horticultural shows in the past. These can be completed by sound installations and other remnants of green art workshops and exhibitions. The cultural skills are challenged by public pianos, which can be found in several public parks of cities in the U.S.—a nice idea that should be considered in Europe and Asia as well.

4.3 Case Studies

This section describes how the above-mentioned planning approaches and green city elements have been applied using concrete examples from China and Germany. Some cities found unique ways to combine several ideas and thus to increase their effectiveness. Of course the examples do not explain comprehensively the situation of the cities, but choose special solutions which are particularly suitable to being explained there.

4.3.1 Beijing: Urban Ecosystem Services as a Guiding Principle

Gaodi Xie and Biao Zhang

Introduction

Beijing, the capital city of China, lies on the northern edge of the North China Plain. It is flanked on the north and west by the Jundu Mountain and Xi Mountain. The

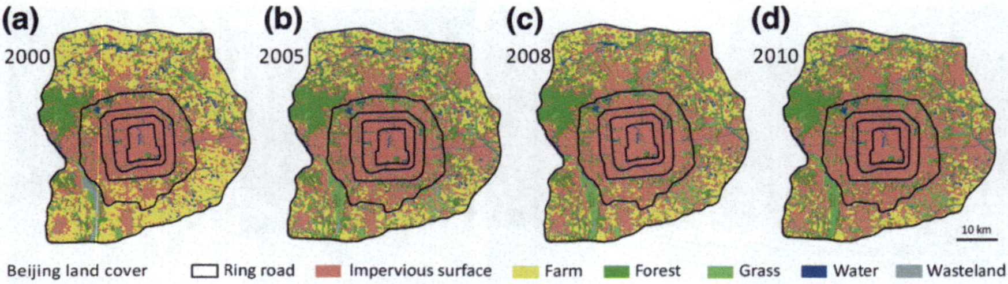

Fig. 4.14 Land cover maps of the study area from 2000 to 2010: **a** the farmland clearly dominated the land cover areas in 2000; **b** the farmland area decreased, whereas the forest and grassland areas increased in 2005; minor changes in land cover were observed between 2008 (**c**) and 2010 (**d**). © Gaodi Xie, Biao Zhang

area administrated by the Beijing municipal government is 16,408 km^2, 62% of which is hilly. The administrative area is composed of two districts (Xicheng and Dongcheng), four suburb districts (Shijingshan, Chaoyang, Haidian and Fengtai) and ten outer suburbs (Daxing, Fangshan, Mentougou, Changping, Yanqing, Huairou, Miyun, Shunyi, Pinggu, and Tongzhou).

Beijing is characterized by a warm temperate continental monsoon climate. The annual average temperature is 11.5 °C, with an average precipitation of 554.5 mm, about 80% of which is concentrated in June–September. In the last decade, extreme climate events and meteorological disasters frequently occurred, such as droughts, floods, and high temperatures. In the context of global warming, the observed changes related to climate in Beijing include an increase in annual average temperatures, possible reductions in average rainfall and runoff, and more heat waves and less frost (Wang 2008).

The population was almost 20.7 million in 2012, and the average population density was 1260 inhabitants/km^2 (Table 2.4). The road network in Beijing consists of ring roads and radial roads as arteries. The road around the Forbidden City is the first ring, and the ring roads beyond this area are the second, third, fourth, fifth, and sixth ones, which were named as such relative to the radial distance of each road from the city center. The urban area of Beijing increased from 183.84 km^2 in 1973 to 1209.97 km^2 in 2005; the built-up area has increased by 1026.13 km^2 over the past 32 years, with an annual expansion of 32.07 km^2 (Mu et al. 2007). Official statistics show that the built-up area of Beijing expanded from 109 km^2 in 1949 to 1350 km^2 in 2009 (National Bureau of Statistics of China 2009).

Urban Green Spaces in Beijing

The total area of Beijing's green space was 61,695 ha in 2009, of which 29.3% was accounted for by public green space, 13% by affiliated green space, 24% by protective forest, 12% by residential green space, and 19.7% by roadside green space; that amounts to a total green space area of 60,472 ha (the remaining 2% consisted of productive green space). The spatial distribution of urban land cover in Beijing is illustrated in Fig. 4.14 based on the data of the RapidEye images in 2000, 2005, 2008, and 2010 (Zhang et al. 2015).

As shown in Fig. 4.14, most of the land cover was concentrated in the core part of the city. Impervious land occupied most of the area, followed by farmlands and forests. The largest net loss from 2000 to 2005 was observed in the farmlands (−295 km^2), whereas the largest net gains were observed in the forests and grasslands. Most of these gains were from the farmlands (Table 4.4). The largest loss between 2005 and 2010 was again in the farmlands. The surface area loss was approximately a quarter of the area lost from 2000 to 2005. The forest area also decreased substantially. Unlike from 2000 to 2005, the largest gain between 2005 and 2010 was observed in impervious surfaces, with a net increase of 142 km^2. Much of the urban growth in the former period was at the expense of the farmlands. Overall, a large area of farmland and water was replaced by impervious surfaces from 2000 to 2010. Grass areas and forests benefitted most from the farmland and wasteland losses.

The changes in urban green spaces and their landscape metrics from 2000 to 2010 are presented in Fig. 4.15. The total area of the urban green spaces in Beijing continuously decreased in the period from 2000 to 2010. The urban green space area amounted to 1041 km^2 in 2000 and only 842 km^2 in 2010, which suggests that almost 20 km^2 of the urban green spaces were replaced by impervious surfaces annually. The Aggregation Index (AI: a landscape metric that measures the tendency of green space elements to be connected with each other) of the urban green spaces constantly decreased over time from 92 in 2000 to 88 in 2010. The Largest Patch Index (LPI: a landscape metric that measures the area of the largest green space element) of the urban green spaces decreased substantially from 7.5 to 5.2 in the period from 2000 to 2005, whereas only a gradually decreasing trend was observed from 2005 to 2010. Thus, the landscape patches of the urban green spaces in Beijing became considerably more isolated and fragmented over the span of 10 years.

Key Ecosystem Services of Urban Green Spaces in Beijing

Ecosystem changes (cf. Fig. 4.14) and ecosystem service assessments of the Beijing municipality were the focus of several studies in recent years. Table 4.5 gives a brief overview of results for key regulating ecosystem services.

Table 4.4 Changes in urban land cover areas in Beijing from 2000 to 2010 (km^2)

Land cover year	Forest	Grass	Farm	Wasteland	Water	Impervious surface	Green space
2000–2005	135.20	91.96	−295.09	2.13	−11.56	77.37	−65.80
2005–2010	−73.40	31.29	−76.35	−14.23	−9.15	141.83	−132.69
2000–2010	61.80	123.25	−371.44	−12.10	−20.71	219.20	−198.49

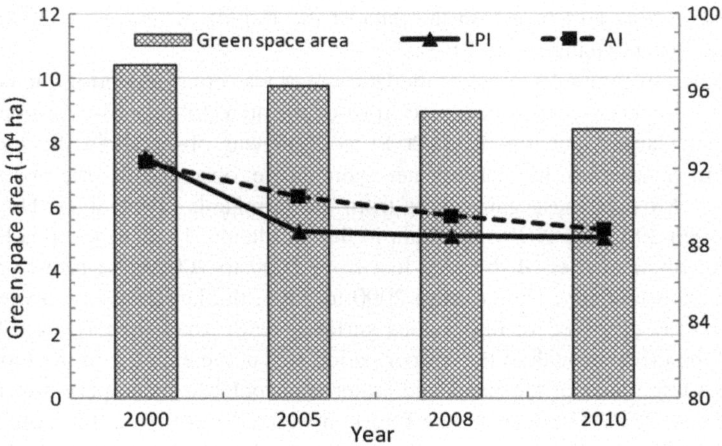

Fig. 4.15 Changes in the area and landscape metrics of the urban green spaces in Beijing. The green space area (bars according left y-axis) in Beijing continuously decreased from 2000 to 2010. Landscape metrics (i.e., LPI and AI, lines according to right y-axis) also showed decreasing trends (modified after Zhang et al. 2015)

Guiding Principles for the Improvement of Urban Green Spaces in Beijing
The present studies (Table 4.5) provided several lessons for policy, practice, and research in developing new urban green spaces. A suitable green space system can provide a variety of ecosystem services; therefore, under the scenario of global climate change and rapid urbanization, clear guidance is needed for local authorities and the public on how best to recognize and manage urban green spaces.

First, the area of urban green space has a clear positive relationship with ecological and environmental benefits, such as cooling effects and storm water runoff regulation. With increasing population in Beijing, more urban green areas have been constructed, especially for the 2008 Olympic Games. Owing to the scarcity of land resources, most of the new green spaces are located outside of the 4th ring road. Beijing municipality realized among others the 'Greenbelt building program' (4th–5th ring with 125 km^2 and 5th–6th ring, 163 km^2), the 'Three North' shelterbelt construction (afforestation of 5188 km^2 in 2012), the 'Country parks plan' (establishment of four new country parks), and the "Million Acres of the Plain Afforestation Action" (planting of more than 50 million trees; afforestation of 650 km^2) (Xiao 2015).

Planning more green spaces in an urban environment can be a real challenge, especially when available lands are scarce and the pressure to develop becomes stronger. In addition, several papers conveyed the importance of the structural characteristics of green land; there are relatively strong correlations between the structural types in each green area and ecosystem services (Sect. 4.2.1). Urban foresters, ecologists, and landscape planners should take advantage of this type of

Table 4.5 Evaluation of selected regulating ecosystem services in Beijing

Ecosystem service	Assessed values
Cooling effect (climate regulation)	The green space of Beijing could absorb 3.33×10^{12} kJ of heat due to evapotranspiration over the entire summer (on average), per hectare the urban green area of the municipality was able to absorb 8.35×10^{8} J/day of heat due to evapotranspiration. This has the potential to eliminate 3.09×10^{8} kWh of electrical power demand for air conditioning and to prevent 243,000 t CO_2 emissions from power plants per year (Zhang et al. 2014; Sect. 3.3.1)
Rainwater runoff reduction (water regulation)	Calculations showed the potential of green space to reduce runoff by 2494 m^3, and a total volume of 154 million cubic meters of rainwater can be stored in urban green spaces, which almost corresponds to the annual water needs of the urban ecological landscape in Beijing. The total economic benefit was estimated at 1.34 billion RMB in 2009, which is equivalent to three-quarters of the maintenance cost of Beijing's green spaces. The value of rainwater runoff reduction was 2,177,000 RMB per hectare (Zhang et al. 2012; Sect. 3.3.2—Example 1)
Air pollution reduction	Based on analyses of satellite images and field surveys to establish the characteristics of current urban forests in the central part of Beijing, the influence of the urban forests on air quality was studied using the Urban Forest Effects Model. The results of Yang et al. (2005) showed that there were 2.4 million trees in the central part of Beijing. The trees in the central part of Beijing removed 1261.4 tons of pollutants from the air in 2002. The air pollutant that was most reduced was PM10 (particulate matter with an aerodynamic diameter below 10 mm); the reduction amounted to 772 tons. The carbon dioxide (CO_2) stored in biomass form by the urban forests amounted to about 0.2 million tons (Sect. 3.3.3)

data to plan, design, and manage green spaces in heat island areas. Policies could be used to encourage the optimal structure and composition of urban green spaces through green space strategies. Urban green spaces should be multifunctional, and their value has to be properly appreciated. We should give adequate attention to urban green spaces that offer important services and establish mechanisms of economic compensation for the people who conserve those spaces. In this way, policies can be employed to encourage the optimal structure and composition of urban green spaces through green space strategies. Thus, climate change and urban sprawl provide opportunities as well as threats for urban green spaces, and city managers should pay more attention to the role of urban green spaces in rainwater regulation and to the scientific management of urban green spaces.

4.3.2 *Berlin: The Revival of Urban Gardening—A Chance for Humans and Nature*

Martina Artmann

The origin of urban gardens in Europe stretches back to the nineteenth century. Their appearance was driven by the undersupply of fresh food for the urban poor population arising in the course of urban industrialization (Barthel et al. 2013; Keshavaraz and Bell 2016). Especially in the periods of the First and Second World Wars, the European gardening movement was incentivized by the need for urban food production due to the wars and economic depression. Cities provided the poorest residents with a piece of land which they could use for their own vegetable production. Food was cultivated in private gardens or vacant lots to compensate for the lack of food production. Due to these circumstances, urban gardening rapidly lost popularity in the postwar period in 1951–1972, since urban gardening was associated with poverty and wartime. However, since the early 1970s urban gardens have been undergoing a revival in Europe. One main reason for this turnaround can be seen in the growing disconnection between humans and nature and the increasing pressure on nature by human activities. Through the rising awareness that enhancing sustainable urbanization cannot work without making the cities greener, urban gardening is gaining an increasing relevance and popularity in Europe again (Keshavaraz and Bell 2016).

Especially Germany has a long urban gardening tradition, which is rooted in the so-called allotment garden movement pushed by the physician Daniel Gottlob Moritz Schreber (1808–1861). He initially had the idea of creating through allotment gardens a place for children to provide them with fresh air and useful occupation. Therefore, allotment gardens are today still often called 'Schreber gardens' in Germany (Drescher 2001). The size of an allotment plot in Germany ranges between 300 and 400 m^2. The allotments are used for noncommercial food cultivation or recreation. The plots often include garden sheds, which are used to store tools and for shelter. Moreover, fruit trees, ornamental flowers, lawns, and vegetable beds can be found (Breuste 2016). According to the German Federal Allotment Garden Law (*Bundeskleingartengesetz*), the allotment plot must be embedded in an allotment garden estate consisting of several allotment plots as well as joint facilities such as a club house or playgrounds. The allotment garden estates in Germany are usually organized in allotment associations. Gardeners have to pay a small membership fee to the association, which pays the owner of the land. The land is mostly owned publicly, privately, or by the church. Within the urban fabric, allotment gardens hold an important share of the recreational areas in Germany. Especially in East Germany, a high proportion of allotment gardens in relation to the city´s territory can be found, such as in Leipzig (3.2%), Halle (3.6%), and Berlin (3.5%) (Breuste and Artmann 2015).

It is argued that no other comparable metropolises have more private gardens within the direct area of the inner city than the German capital and the biggest city

Fig. 4.16 Allotment gardens in Berlin (*left*) and Dresden (*right*). © Robert Bendner

in Germany, Berlin (3.47 million inhabitants in 2014): Berlin has around 3000 ha of allotments and 915 allotment estates. Three-quarters of the allotments are owned by the federal state of Berlin. The sizes of the allotment estates vary between 400 and over 491,000 m^2, including between 121 and 10,294 allotment plots per allotment estate (Senatsverwaltung für Stadtentwicklung und Umwelt n.y.). The importance of the allotments within the urban fabric is emphasized by the Berlin allotment garden development plan (Senatsverwaltung für Stadtentwicklung 2004). It is stated that allotments are an important part of the green infrastructure, and along with their embedding in the green network, which should be fostered, they contribute to urban climate mitigation (Figs. 4.16 and 4.31). Since the maintenance of the allotments requires involvement of civil society, allotments promote social activities and cohesion. Thus, allotment gardens can contribute to a decrease in social anonymity and individualization within metropolitan areas. To increase the value of allotments for urban recreation, the plan recommends opening the allotment estates to the community so that especially residents living near the allotment sites can use their seating and playground areas (Senatsverwaltung für Stadtentwicklung 2004). Current studies in Austria suggest that the recreational use of allotments increased in recent years. Recreation but also nature experience is now more important to the allotment gardeners than food supply. The allotment gardeners also use their allotment sites to bring younger family members in touch with nature, for instance to observe wild animals (Breuste and Artmann 2015).

In general it is assumed that the typical allotment gardeners are pensioners who spend the majority of their free time in the allotment garden but also share it with other family members (Breuste 2016). However, along with the increasing interest in urban allotment gardening, the allotment gardeners also became younger, as examples in Berlin show. According to the State Association of Garden Friends Berlin (*Landesverband der Berliner Gartenfreunde*), the average age of the allotment gardeners decreased during the last ten years from 67 years to now 56.5 years. Today, in particular young families apply for allotment gardens to provide green spaces to their children (BZ Berlin 2016). However, in Berlin there are long waiting lists to get an allotment garden. In 2013, 10,000 residents were waiting for an allotment garden, and in 2015 already 14,400 applicants were on the waiting lists. However, only 3106 allotments became available in 2015 (BZ Berlin 2016).

Besides the classical allotment garden, cooperative forms of urban gardening developed within the last years in Europe, including in Germany. New gardening communities arise through their motivation to reclaim public spaces. Also the names of the gardening initiatives mirror their focus on the community. The names they take refer to community gardens, neighbourhood gardens, or intercultural gardens. Studies across Europe showed that the community gardeners are very heterogeneous with respect to their age, gender, education, or gardening experiences (Ioannou et al. 2016). However, their common focus is on empowerment, solidarity, and human well-being while taking care of the garden from its conceptualization till its creation and maintenance (Ioannou et al. 2016; Rosol 2012).

Also in Berlin, a range of collective garden forms can be found. For instance, there are around 74 intercultural gardens (www.anstiftung.de/berlin) with different foci such as on intergenerational contact. One example of this is the Berolina garden of generations in Berlin (*Berolina-Generationengarten*). The garden was planned and developed and is being maintained by elderly people together with two neighboring daycare facilities for children. Other intergenerational gardens also exist, which provide for instance raised beds for elderly people, handicap toilets, parties and food markets to raise awareness about the projects. Besides intergenerational contacts, the community gardens in Berlin also promote cultural and personal exchange between residents and specific population groups such as refugees, disabled people, mentally ill persons, unemployed persons, and single parents with children and families. The gardens are therefore mostly oriented toward fostering a common experience of nature, learning about ecological gardening, social commitment, and successful integration. According to Rosol (2012), the majority of the community gardens in Berlin cultivate only shrubs, flowers and occasionally trees, providing benefits in terms of environmental qualities, aesthetics, and recreation rather than food supply. Since space for gardening is rare in cities, roofs are also used for community gardens such as the cultural rooftop garden *Klunkerkranich* in Berlin (http://www.klunkerkranich.de/). By restoring the top of a car park, the community garden aims to make the city greener and to create a place for urban participation dealing with a range of topics around urban ecology, such as urban gardening and permaculture, biodiversity and sustainability (see Fig. 4.17).

Along with the shift in lifestyle by modern society, it is to be expected that allotment gardens as well as new garden forms such as community gardens will become even more important in European cities and worldwide. Reflecting the benefits urban gardens provide to the residents and the urban ecosystem, urban decision-makers and planners need to consider their value when it comes to urban development. Even in densely built-up parts within a city, sufficient space for urban gardens should be provided, such as in gaps between buildings, in brownfields, or on rooftops.

Fig. 4.17 The community garden Klunkerkranich on the top of a car park in Berlin. © Martina Artmann

Table 4.6 Urban greening of Shanghai (Shanghai Municipal Statistics Bureau 2006, 2014, in Breuste et al. 2016), in ha

Year	Urban green space[a]	Public urban green[b]	Parks	Urban forest	Green coverage rate in %
1990	3570	983	712		12.4
2000	12,601	4812	1153		22.2
2005	28,856	12,038	1521	1284	37.0
2010	120,148	16,053	–	83,340	38.2
2013	124,295	17,142	–	84,152	38.4

[a]Including roadside green, afforestation, nurseries, parks, not all accessible and public
[b]Including all parks and other publicly accessible green, but no roadside green and forests

4.3.3 Shanghai: A Gray City Becomes Green, Developing with Green Infrastructure

Jürgen Breuste

Since the 1990s, the megacity Shanghai has been developing extraordinary dynamics; old built-up structures have been replaced by new ones, which extend into the surroundings of the original city (Table 4.6). About 24 million people live in the city area of 6341 km^2 (LGSMP 2015)—this makes Shanghai about seven times bigger than Berlin by population and area. The task in Shanghai is to develop a resilient and suitable urban structure (Shanghai Municipal Statistics Bureau 2006,

Fig. 4.18 Green Space System Plan. Source: SUPAB, SUPDRI n.y.

Liu and Huang 2007, SUPAB), mostly with compact built-up building blocks and limited urban sprawl, based on existing compact urban structures with only very little urban green. In 1978, urban green in Shanghai had only 8.2% share of the urban area; the public urban green per capita was only 0.35 m^2 (Shanghai Municipal Statistics Bureau 2006, 2014; Breuste et al. 2016).

The current vision of the Shanghai Master Plan 2015–2040 is "striving for the excellent global city."

The detailed objectives are:

- Global Competitiveness: a City of Prosperity and Innovation
- Sustainable Development: a City of Green and Well-being
- Cultural Attractiveness: a City of Happiness and Humanity (LGSMP 2015).

In 2003, the urban planners decided to enhance the urban development with an ambitious Greening Master plan consisting of:

- two green rings, including afforestation areas, tree nurseries and recreational parks providing ecological and economic functions,
- eight big connected green islands ('green lands') around the city to improve the urban climate,
- green corridors along the main roads, railway lines, and water streams into the city, and
- public green spaces accesible within a distance of 500 m for all inhabitants of the residential areas (Leung 2005; Breuste et al. 2016) (see Fig. 4.18).

The declared goals were the protection of biodiversity, climate mitigation, and protection of drinking water resources, together with provision of recreational sites. Since 2000, the public green space per capita has been doubled every 2 years; eight million square meters of new public green space have been developed (You 2009). Based on these numbers, Shanghai was recognized by the Ministry of Civil Affairs as a 'National Garden City' in 2004. In 2010, the green coverage rate reached nearly 40%, with nearly 10 square meters of green per capita and a forest coverage rate of 25–30%. The urban forest along the roads and big forest areas outside the built-up city make up about two-thirds of all the urban green. The urban policy "Wherever there is a road, there is greening" resulted in 9.9 million newly planted urban trees in 2013 (Liu and Huang 2007; Li et al. 2008; Shanghai Municipal Statistics Bureau 2014).

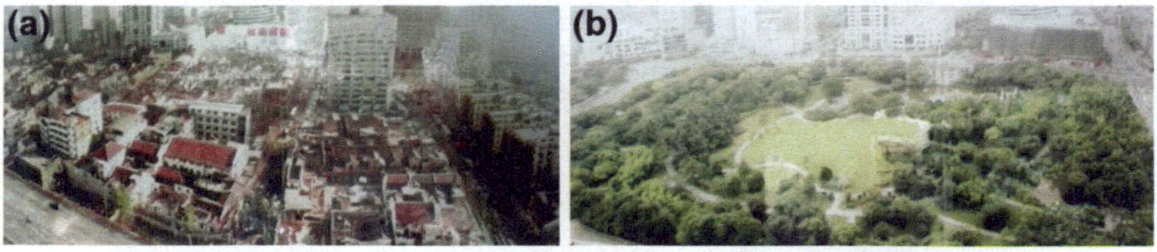

Fig. 4.19 Site of urban park Yangzhong Greenery, Shanghai, 2000 (*left* before, *right* after, info panel on site). Photos: J. Breuste

Despite this dynamic urban greening development, the share of urban parks is still not sufficient. The biggest urban park is the 140 ha Century Park, developed in the new city extension district Pudong. There is still a tremendous need for realizing accessible urban residential parks in old and new residential areas with high population density.

In the densely built-up areas, new green space could only be established by replacing old built-up structures. An example is the Yangzhong Greenery in the Huangpu section of the Central City District. The new urban park of 11.85 ha was established in 2001 on a former residential built-up site after removing 4837 families and pulling down the old built-up stock (Fig. 4.19).

Another project was to build China's flagship Eco-City Dongtan, planned for up to 400,000 inhabitants (by 2050), started in 2005 on a 6.5-km^2 start-up area on Shanghai's Chongming Island. The entire island is planned to become a 'green island' of high-quality green infrastructure and nature protection sites (Chongming City 2003; Liu and Huang 2007). Dongtan was planned by the city of Shanghai as one of nine towns to relieve overcrowding in the central city and to offer exemplary, well-developed environmental conditions for its inhabitants. The project aimed for reduction in energy demand, supply from bioenergy, renewable energy in buildings, reduction of landfill waste, and almost zero carbon emissions (Castle 2008; You 2009). The first demonstration phase was to have been completed in 2010 for the World Expo but could not be realized until now.

Other bits and pieces of urban green infrastructure are visible in selected realized building projects, like Hongqiao Low-Carbon Business District. The Low-Carbon Business Center (1.4 km^2) defines ambitious targets in building technologies, traffic planning, energy production, and green infrastructure, including greening of buildings in a densely built-up area (SBA design 2013) (see Fig. 4.20).

The new Master Plan of Shanghai 2015 aims to develop Shanghai as a 'Green and Low-Carbon Ecological Environment' (LGSMP 2015) to

- consolidate the ecological spatial structure (by protection of the marine and continental ecological basis, construction of a coherent ecological spatial

Fig. 4.20 Hongqiao low-carbon business district. Source: SBA design 2013

system, building ecological nodal spaces, implementing very strict basic farmland protection regulations, and perfection of the policy mechanisms of ecological protection),

- enhance low-carbon and energy-efficient development (by reduction of carbon emissions, optimization of the energy structure, lowering of energy costs of industry and buildings, and encouraging green and low-carbon transportation), and
- strengthen environmental protection and comprehensive regulation (by improving air and water quality, soil preservation and pollution treatment, and enhancing solid waste disposal).

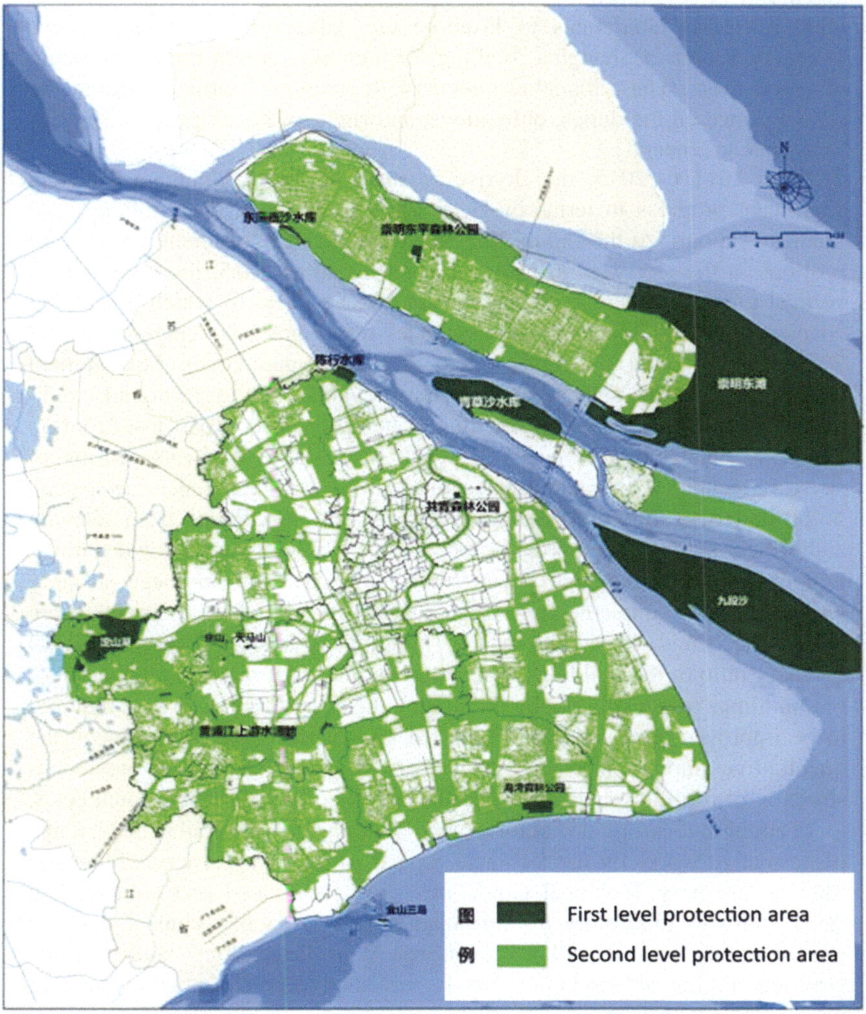

Fig. 4.21 Ecological network in Shanghai LGSMP. Source: 2015

With this new Master Plan, a new phase of ecological development is envisaged. It includes the idea of increasing ecological services on the existing green spaces. Since the space for further extension of urban green infrastructure is limited, the target is to improve and to protect the existing urban green infrastructure network and its provision of ecological services in the context of urban growth and redensification. This will still be a task for the next decade at least (see Fig. 4.21).

Box 4.5 Skyrise Program in Shanghai (*by Nannan Dong, Yuelai Liu, Song Liu*)

As early as 2007, Shanghai started to encourage the development of ‘skyrise greenery’, i.e. the greening of roofs and facades. In 2014, the local government released “Guidelines to Promote the Advancement of the Skyrise Greenery Sector in Shanghai,” calling for increased efforts for development of green roofs. The “Shanghai Greenery Regulations” further clarified the state’s stance on the duties, obligations, and rights of developers with regard to skyrise greenery.

From 2011 to 2015, the skyrise greenery sector in Shanghai achieved significant progress in terms of green coverage area, quality, policies, and financial support. At the end of 2009, the estimated area of green coverage on rooftops in the city was 90 hectares. By the end of 2015, the total area of vertical green coverage in Shanghai was 262 hectares, of which green roof coverage was 218 hectares. This translates into a green coverage of around 7% of all potential rooftop surfaces. The Shanghai Green Management Bureau used remote sensing approaches to evaluate the location and conditions of those rooftoops with potential of being greened. Apart from rooftop greenery, Shanghai also encourages the installment of diverse forms of skyrise greenery. This rapid advancement in the skyrise greenery sector can be attributed to the enforcement of policies and regulations from relevant authorities. During the approval procedure of new construction and renovation projects, the Shanghai Green Management Bureau has been strongly encouraging the implementation of vertical greenery. Moreover, there are relevant laws and regulations for new buildings to achieve a certain vertical greenery ratio of the buildings. For example, in the approval of renovation of old buildings (especially public buildings), the share of green area is required to be improved. Considering the limited size of the renovation project, the quantitative improvement of greenery usually depends on the new construction of rooftop green, although this incurs considerable additional costs.

In addition to strong law enforcement, Shanghai has also published a range of normative works on the technical aspects of skyrise greenery. “Shanghai Skyrise Greenery Technical Guidelines” was formulated and published in 2008, followed closely by the formulation of a series of technical standard guidelines, such as “Handbook on Green Wall Technology” and “Skyrise Greenery Technical Standards.” Newly published technical standards in 2015 include “Guidelines on Construction and Management of Green Pillars along

Highways" and "Guidelines on Skyrise Greenery in New Developments"; they are a further indication of an increase in political support toward the skyrise greenery sector in Shanghai.

4.3.4 Munich: Linking Gray and Green Infrastructure

Martina Artmann

The city of Munich is situated in the south of Germany. Munich is the capital of the Free State of Bavaria located in the alpine foothills, with the river Isar flowing through the city from southeast to northeast. In the south of Munich there is a range of forests. Three large conservation areas and large marshland sites are located in the north of the city. 15.4% of Munich's territory is declared as landscape protection areas, and 2.3% is protected as conservation areas (City of Munich 2016). Munich is Germany's third largest city, with around 1.5 million residents. It is also one of the most dense cities, having a population density of 48 residents/ha on a size of 311 km^2 (City of Munich 2015a). Munich can be characterized as a growing city, regarding both population and settlement area. The population is projected to increase from 1.5 million today to over 1.7 million inhabitants by 2030 (City of Munich 2015b). Munich is therefore under pressure to provide new and affordable housing for the increasing number of residents.

To take action against the housing shortage, Munich plans to issue 3500 new permits for housing units and to realize 1800 subsidized housing units per year (City of Munich 2012). However, Munich already today is one of the most sealed cities in Germany. Open and green spaces are a scarce resource and are intensively used. In 2006, 43% of the territory of Munich was sealed, and the proportion increased to 47% in 2012 (www.ioer-monitor.de). In 2013, the urban green supply (recreational areas, forests, blue infrastructure and agricultural areas) was 77 m^2 per capita, which is the lowest supply among big cities in Germany. Due to the growing population, the green supply will decrease in future, and without the development of new open spaces it will be 67 m^2 per capita by 2030 (Becker et al. 2015). Therefore, due to the need to build new housing and related infrastructure through densification, restructuring and development of agricultural areas, the green infrastructure in Munich is under threat.

To achieve a balance between needed further development of residential housing and a sufficient supply of green infrastructure, the city of Munich places its urban development under the *Leitbild* 'compact/urban/green'. That the city's authorities are aware of the growing conflicts between densification and urban green infrastructure can be derived from the fact that the city of Munich is currently conceptualizing the strategy "Long-term open space development 2030" (German: *Langfristige Freiraumentwicklung* 2030). In light of the increasing challenge to

Fig. 4.22 More green in Munich—award for child-friendly residential environment: 'Activity playing garden' Lindwurmstr. 4. © Jutta Polte-Giessel

secure a high living quality in an ever more densely populated city, the concept demands an expanded understanding of open space. Open space must be regarded not only as space devoid of development but as a space with multifunctional or temporary uses. In this regard, Munich aims to maintain and strengthen the multidimensional functions of open and green spaces in terms of social, ecological, and economic values. Green spaces should provide different ecosystem services, such as recreation, physical activities, biodiversity protection, water and climate regulation. The concept points out an urgent need to consider that also parts of the gray infrastructure (e.g., residential, commercial, transport areas) need to perform tasks of open spaces due to limited space resources. Moreover, sustainable protection and development of green spaces requires the involvement of residents to offer them the possibility to actively shape the city and assume responsibility for people and the environment they live in (Becker et al. 2015).

Munich provides a range of opportunities for participative urban green infrastructure development and for integrating the green into the gray infrastructure. One best practice example is the competition "More green for Munich" (German: *Mehr Grün für München*). The competition is biennial and honors exemplary greening projects in six categories: front gardens, yards, outdoor facilities, personal commitment, child-friendly living environment and greening of commercial areas. House owners, tenants, and holders of commercial enterprises are invited to green roofs and facades, to unseal commercial areas, to develop diverse planted yards and biotopes in the inner city or to design natural playgrounds. Through making such citizen greening initiatives visible, residents can act as a role model for society in engaging for ecological and social urban development. An evaluation panel, comprising members of the city council and experts such as environment educators and landscape architects, honors the projects with a certificate and small cash prizes during a solemn reception in the city hall. In 2015 the competition had its fortieth anniversary, and on this occasion Munich's Deputy Mayor Josef Schmid argued that Munich's competition "became a trademark for initiative and responsibility of the residents to promote the garden culture in our city" (City of Munich 2015c, 5). Since 1945, more than 1000 award winners out of 2000 participants were recognized. Figure 4.22 shows an example of a yard that was rewarded in the framework of the greening project. More pictures of project examples can be found online (City of Munich 2015c).

Besides the competition, the city of Munich provides a range of subsidies to interlink green and gray infrastructure, such as for greening roofs of commercial or residential areas. The funding is provided for the extensive greening of previously nongreened roofs, provided that the greening is done on a voluntary basis and not in the course of discharging an obligation under public law. The green roof must have at least a thickness of 8 cm. If a thickness of 10 cm is implemented, an additional reduction of costs can be achieved through a decrease of the sewage charge. In Germany, the split waste water fee refers to the separate charging for sewage and rainwater, with the amount based on the proportion of sealed surfaces of a property. Thus, the fee is a monetary incentive for unsealing and greening of buildings to increase the natural surface water infiltration. To further promote unsealing actions, the city of Munich also provides subsidies for the voluntary unsealing and greening of private courtyards within the program "Green courtyards–green walls" (German: *Grüne Höfe - Grüne Wände*). The focus of the program is on the greening, unsealing, or modification of pavements in favor of semi-permeable surfaces (e.g., grass pavers, cobblestones) of courtyards.

4.3.5 Xuzhou: City in Structural Change

Tinghao Hu and Pingjia Luo

Introduction

Xuzhou (117°20′E, 34°26′N) is located in a water-rich area (Fig. 4.23) of northwest Jiangsu Province, China. It is an important coal-based city with an old industrial base, as well as the only energy production area in Jiangsu Province. There are more than one thousand state-owned enterprises in sectors like coal, machinery, chemical industry, building materials, metallurgy, etc., in Xuzhou, which formed a heavy industrial system based on raw materials and equipment manufacturing industry. However, with the past mining activities, coal resources were depleted, and the industrial development resulted in serious environmental pollution problems and a large number of unrestored coal mining subsidence areas. As a result, transformation became the only route to be implemented. In 2007, the State Council of China issued documents to promote the transformation of coal-based cities. In

Fig. 4.23 Aerial view of Xuzhou (Source: Local Chronicles of Jiangsu Province. Source: http://www.jssdfz.gov.cn/index.php?m=content&c=index&a=show&catid=9&id=15)

fact, transformation work has been carried out in Xuzhou since the end of the 1990s, especially in ecological aspects. This chapter mainly introduces the efforts that have been undertaken in promoting urban ecology in Xuzhou, China, so as to provide a reference for constructing green cities, especially for coal-based cities.

Objective Driving Factors of Structural Change in Xuzhou

Coal Resources Exhausted and Transformation of the City's Orientation

Coal mining activities have a long history in Xuzhou, which can be traced back 900 years, beginning with the Northern Song Dynasty. In 1882, the Liguo Iron and Coal Mining Bureau was founded, and this can be regarded as the beginning of large-scale coal exploitation and processing as well as the beginning of the modern industrial era in Xuzhou (Chang et al. 2011). As the city was an important coal resource base and transportation junction, the industry was greatly supported by the country and gradually formed a heavy industry agglomeration based on coal, steel, calcium carbide, and engineering machinery. Meanwhile, extractive industries were developing rapidly (Shen et al. 2012). At the end of 1990s, the coal resources in Xuzhou were almost exhausted. At the same time, the traditional industrial system was shocked by the rapid development of the manufacturing, commercial and service industry. Thus, the old industry system, environmental pollution and ecological destruction forced the city to face structural change and transformation.

Rapid Urbanization and Urban Sprawl

As in other cities in China, the urbanization process in Xuzhou entered an accelerating period, which manifested in large-scale urban expansion toward the south and east (Chang et al. 2011; Qu et al. 2009). In the past 30 years (1984–2014), construction land increased from 311 to 756 km^2 (Fig. 4.24). The area of water bodies increased by 68 km^2. There is no doubt that the increase in construction land

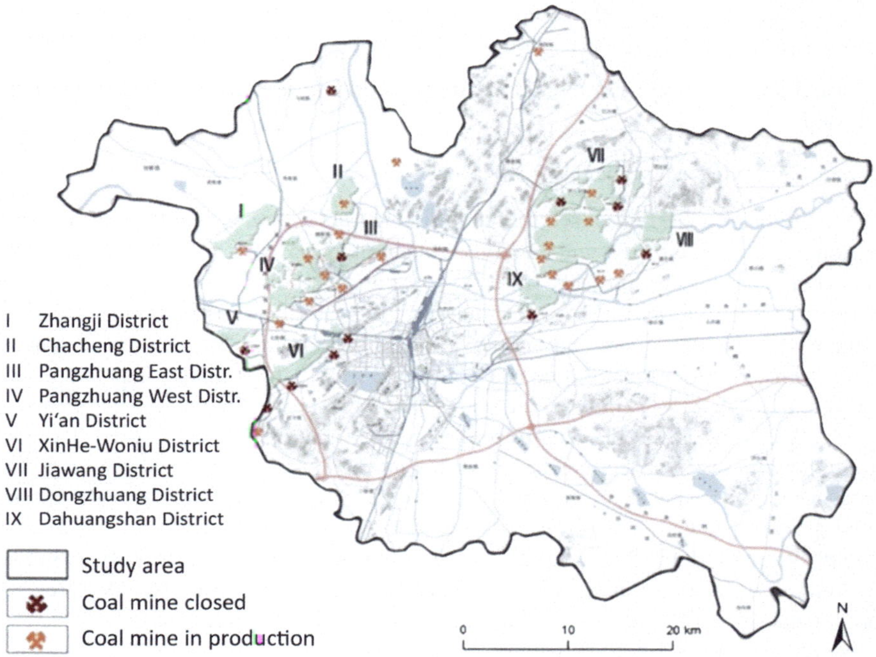

Fig. 4.24 Distribution of coal mining subsidence areas in Xuzhou's urban area. © Thinghao Hu

is due to the process of urbanization, while the increase in water bodies is mainly caused by the coal mining subsidence areas in Jiawang and Tongshan district. As a result, the extent of urban construction land already lies outside the range of the definition in the Xuzhou Master Plan for 2020. Problems like the rapid growth of the urban population, environmental degradation, urban sprawl and land resource pressures appeared, and especially the conflict between land resources and ecological security.

Large Number of Unrestored Coal Mining Subsidence Areas

Coal mining activities have brought considerable economic benefits to the city, while resulting in stumbling blocks for the city's development. Subsidence areas caused by coal mining activities not only directly affected various production activities, but also had adverse effects on the urban ecological environment (Chang and Koetter 2005; Chang and Feng 2008). According to the data from 2012, 16,133 ha of coal mining subsidence areas were found in the Xuzhou metropolitan area, of which 72% remained unsteady and unrestored. These subsidence areas are mainly distributed in the Tongshan district, the Jiangwang district, and Pei County.

28 towns and more than 90 administrative villages were affected to various degrees (Feng et al. 2016). The distribution of subsidence areas is shown in Fig. 4.24.

Toward a Green City: Ecological Strategies and Implementations of Structural Change

Urban Gardening Movements in Xuzhou

In the 1990s, Xuzhou proposed the goal to construct a modern ecological city and regional commercial city with an investment of nearly 10 billion yuan in construction. The flagship projects "three squares and one road" and "three roads and one river" were carried out. "Three squares" refers to the construction of Huaihai Square, Pengcheng Square and People's Square, which provided new green spaces for citizens. In the project "three roads and one river," three main roads that cross the urban area and a Yellow River scenic belt were reconstructed and regreened.

In 2002, Xuzhou set the goal to become a "national garden city." After 3 years, the indicators green space in urban built-up area, green coverage, and public green area per capita respectively reached 3566.57 ha, 36.86% and 7.82 m^2. The urban road greening rate is 100%. The city basically formed ecological patterns by the composition of dots, lines, flakes, noodles, meshes, rings. It generally exhibits mountains as a skeleton, rivers and roads as a network, parks and squares as ornaments. In 2005, Xuzhou attained the title of "national garden city."

From 2001 to 2010, 16,921 ha of green spaces were newly formed. By the end of 2010, the green cover percentage increased to 42% in the urban area, and the forest coverage rate reached 31.3%, ranking first in Jiangsu province (Xuzhou Statistical Yearbook 2015). The city regained its characteristics of "one city of green and half a city with a lake" and transformed form 'gray' to 'green'. The city's appearance, function, and environmental quality were significantly improved and enhanced.

From 2010 to 2014, 800 million yuan were invested in urban ecological construction. The "Troop into Barren Hills" and "More Green Space" projects were important actions launched. During this period, 6900 ha of afforestation and 120 landscaping projects were realized. The green space area in urban built-up areas increased to 1168 ha, including 610 ha of green park areas. The two indicators "proportion of green space in built-up area" and "public green area per capita" increased to 1.96% and 1.51 m^2, respectively, in the past four years (Fig. 4.25). The green coverage rate in built-up areas rose from seventh place in 2010 to second place in 2014 in Jiangsu Province. In 2016, China's Ministry of Housing and Urban–Rural Development announced the list of the first 7 national ecological garden cities: Xuhzou, Suzhou, Kunshan, Shouguang, Zhuhai, Nanning, and Baoji. Xuzhou ranks first in this list.

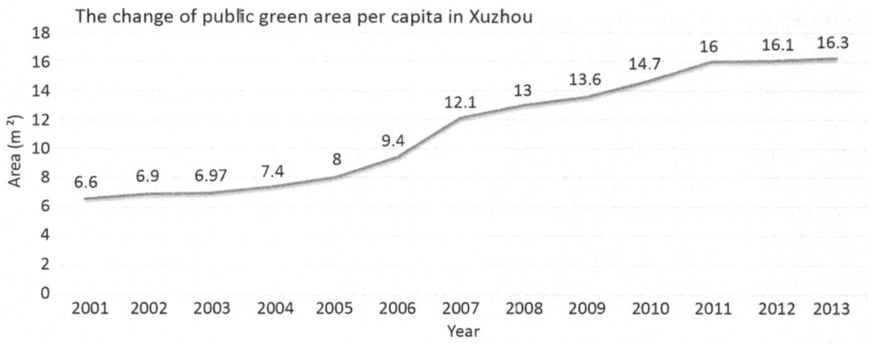

Fig. 4.25 The change in public green area per capita in Xuzhou. *Source* Xuzhou statistical yearbook 2014

Ecological Restoration of Coal Mining Subsidence Areas

In 2008, Jiangsu's provincial government initiated the revitalization of old industrial bases in Xuzhou. The next year, Chinese Premier Wen Jiabao and German Chancellor Angela Merkel jointly attended the signing ceremony of an ecological flagship cooperation project in Xuzhou between Jiangsu Province and North Rhine-Westphalia's government in the German chancellery. "Implementation of an ecological restoration project for coal mining subsidence areas in northern Xuzhou" was one of three main components of this project (Lin et al. 2009). It was against this background that the project "Ecological Restoration of Mining Subsidence in Xuzhou" emerged. This project is a beneficial supplement to the overall urban planning process. It covers all coal mining subsidence areas in the Xuzhou area. This project consists of three levels of restoration measurements.

- General plan for coal mining subsidence areas in Xuzhou's urban areas
- New plan for Jiuli Lake and Liuxin experimental area
- Spatial plan in Jiuli Lake area.

This ecological restoration plan for coal mining subsidence areas in Xuzhou is a comprehensive combination of environmental restoration, urban sprawl control, village relocation, and new rural construction. In the meantime, a cooperation platform was established for different interest groups to increase their awareness of social, economic, and ecological aspects. This plan is a powerful complement to urban master planning and emphasizes the ecological use of coal mining subsidence areas. By remodeling and renovating coal mining subsidence areas, the potential value of abandoned lands could be fully developed as an important resource.

Planning Approaches Toward Green Cities

Conservation and Creation of Green Infrastructure

Because of the pressure of social and economic development toward ecological resources, a series of efforts were undertaken in the area of urban planning. In this section, some ecological protection planning documents are summarized and interpreted, in order to provide a reference for the construction of green cities.

In order to provide a sound guide for construction and management of urban green space, the city compiled the **Xuzhou Urban Green System Plan** (2005–2020) on the basis of the Xuzhou Master Plan (2005 edition). In this plan, a green system of "polycyclic belts, mosaic patches and multiple corridors" was formed. In the built-up area, the green space was planned as "one circle, four belts, six rings, 19 corridors and 14 cores."

For the purpose of fulfilling the requirements of protecting key ecological function areas, Xuzhou compiled the **Xuzhou Key Ecological Function Areas Plan** (2011–2020). This plan covers the metropolitan area. The following three aspects were included in this plan:

- It defined types of key ecological areas and the main functions of ecological services.
- It proposed a list of key ecological function areas, its scope, and management requirements.
- It proposed regulatory and protection requirements for various types of areas. In a total of 11 classes, 50 key ecological function areas were included, with an area of 2595 km^2.

Blue Space Protection

Wetlands are important elements of urban ecological systems. Xuzhou was surrounded by a large number of subsidence areas caused by coal mining activities. These areas are not suitable for urban construction. On the other hand, these areas have great potential to be restored as wetlands (Chang and Feng 2008). The Xuzhou Environmental Protection Bureau therefore compiled the *Xuzhou Wetland Conservation Plan (2011–2020)*. The goal of this plan is comprehensive maintenance of natural ecological features and of basic functions of wetland ecosystems. This plan helps to repair ecological elements and maintain basic functions of the wetland ecosystem, so as to bring the ecosystem into a stable condition. Meanwhile, through projects demonstrating sustainable utilization of wetland resources, a framework for a wetland regulation system was set up, including wetland monitoring and publicity.

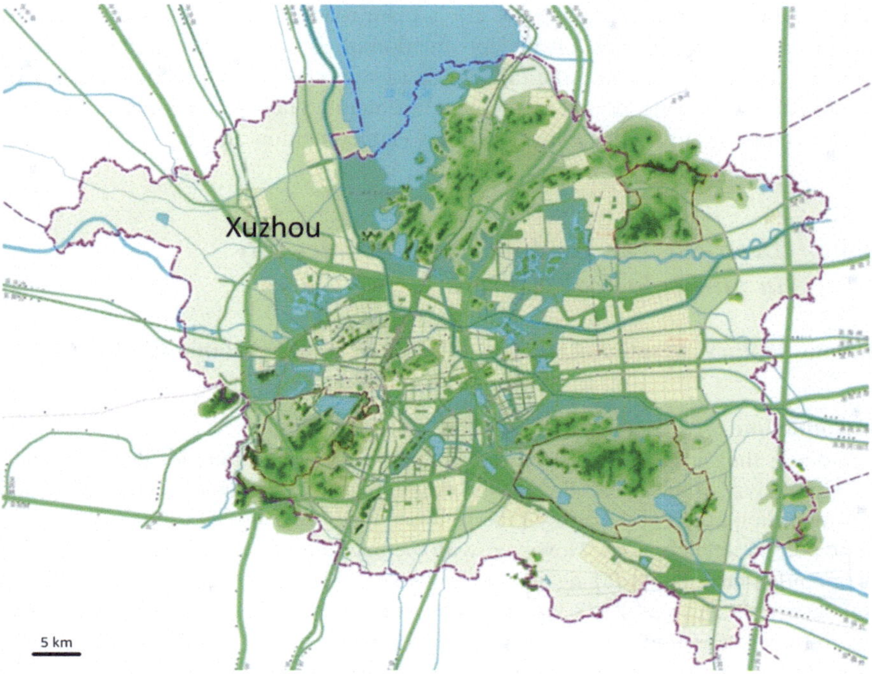

Fig. 4.26 Green space system in Xuzhou's urban area (Source: The Planning of Xuzhou Metropolitan Area)

Biodiversity Protection

In 2011, Xuzhou's municipal government compiled the *Xuzhou biodiversity conservation plan (2011–2020*, Fig. 4.26). This plan takes zonal vegetation, wetland resources, rare wild animals, and famous wood species as a protection priority, with repairing local biological habitats as the key, in order to set up and improve biodiversity conservation networks made up of nature reserves, forest parks, scenic spots, urban green spaces, famous tree species, lakes, wetlands, farmland, etc. The ultimate goal of this plan is to realize multilayer biodiversity protection and sustainable use of biological resources.

Air Quality and Urban Heat Island Control

In order to carry out the relevant contents and requirements in the Xuzhou Master Plan and enhance the ability to control urban self-purification, the microclimate as well as the heat island effect, Xuzhou compiled the *Wind Corridor Plan of Xuzhou* in 2014. In this plan, the framework was built from two aspects. The first is wind corridor construction. According to the dominant wind direction, it stipulates a certain width of greenbelt along the city's main roads, highways, railways, rivers, and lakes. The second aspect is the construction of an urban forest greenbelt circle, so as to direct the wind into the urban area to stop and purify polluted air. By the combination of a corridor and a greenbelt, air quality and urban heat island control can be improved to a great extent, owing to superposition effects.

4.3.6 Dresden: The Leitbild *of a Compact City in an Ecological Network*

Juliane Mathey, Martina Artmann, Olaf Bastian and Stefanie Rößler

Introduction

Dresden is the capital city of the German federal state of Saxony located in the east of Germany. After a decade of shrinkage due to decreasing birth rates and out-migration after the collapse of the state-planned economy, Dresden experienced large population decreases. However, with the beginning of the twenty-first century, the population is increasing again, and in the year 2016 553,036 residents are living in Dresden on an area of 328 km^2 (LH Dresden 2016b). The territory stretches over several landscape units. The city is embedded between the foothills of the Eastern Ore Mountains, the Lusatian granite slab and the Elbe Sandstone Mountains. The landscape of Dresden is shaped by the river Elbe with its broad floodplains mostly covered by semi-natural meadows crossing the entire city from southeast to northwest. The inner city is greened by parks (Fig. 4.13) and avenues (Fig. 4.29 *right*) but also by many small habitats of partly rare and threatened plant and animal species. In the northern and eastern fringe Dresden has large areas of forests. The north and west of the city are predominantly agricultural and forest areas. In total, with 62% of forest and green areas Dresden has a high share of green and recreational spaces, including for instance nature and landscape conservation areas, parks, allotments, and cemeteries (LH Dresden 2014).

To react to increasing urbanization and to minimize urban sprawl, the '*Leitbild*' of a 'compact city' is stated in the German National Sustainability Strategy (see Sect. 2.2.2). Urban green spaces with their manifold functions are vital elements of the urban fabric. In particular, densification processes require a strategic approach that considers trade-offs between infill development and the protection and development of green spaces, resulting in environmental dilemmas and more complex conflicts (De Roo 2000).

In order to supply ecosystem services and urban biodiversity, in particular, it is therefore necessary to develop an integrated system of urban green and open spaces that takes account of all elements of green infrastructure. This ensures a useful climatic adaptation effect, especially when a dense network of green sites is in place. The green space system plays a vital role in supporting the habitat network in a city and its connectivity to the surrounding landscape, as well as in providing recreational areas that can be easily accessed by local residents. To increase the green space supply, the green infrastructure concept (see Sect. 2.2) suggests a physical or functional integration of urban green along with the planning of gray infrastructure, for instance when planning transport networks or vacant plots (Davies et al. 2015; Kambites and Owen 2006; Keeley et al. 2013).

But for urban developers and planners it is not easy to face the challenge of creating compact and green cities simultaneously. If a strategy for the development of the entire system of urban green space is in place, which also takes account of priority areas for infill development, then this can serve as a firm basis for decision-making (Hansen and Pauleit 2014).

Urban Green Spaces in Dresden and Current Challenges
The current challenges to urban development are interrelated with past and future re-urbanization processes. Increasing population in connection with housing requirements leads to pressure on undeveloped and open spaces (LH Dresden 2016a).

Guiding Principles for Dresden and Its Urban Green Spaces
The focus of the urban development of Dresden is thereby placed on urban infill development and the activation of housing development areas in the inner city and the revitalization of brownfields dating to former periods of shrinkage, in order to counteract processes of urban sprawl. Although in the past the extent of green spaces increased (mostly through renaturation of brownfields), the city of Dresden states that along with future urban development the protection and expansion of green spaces needs to be particularly highlighted, given their increasing importance for urban sustainable development (LH Dresden 2016a).

Dresden developed many approaches and strategies to manage challenges like adaptation to climate change (heat, drought, flooding), securing ecosystem services, stopping the loss of biodiversity, and solving brownfield problems. Examples are:

- IRKAP (Integrated Regional Climate Adaptation Program me): Formulates the main requirements of adaptation to climate change and aims to create green and compact cities (REGKLAM 2012, Box 4.6).
- Green Space Standards for Dresden: Standards for the quantity and quality of all urban green spaces in Dresden are fixed. The different kinds of urban green spaces as well as all occurring building types are considered.
- Dresden Landscape Plan: Aims at ensuring sustainable land management and the potential utilization and regeneration of main natural assets (soil, climate, water, air, species and biotopes, landscape, humans), taking into consideration their landscape functions.

The "Compact City in an Ecological Network"—'Leitbild' for Urban Development in Dresden

Dresden's landscape plan, which addresses the entire city area, aims for sustainable spatial development by promoting a compact city permeated by a green network. In order to consider the demands both to foster a compact city and to maintain and develop the green infrastructure, the current draft of the landscape plan of the city of Dresden is placed under the '*Leitbild*' (overall concept) "Dresden—the compact city in an ecological network" (German: "Dresden—die kompakte Stadt im ökologischen Netz") as a guiding principle for the entire city development (LH Dresden 2014). The idea is to achieve a spatial concentration of urban functions by means of a compact, cell-like structure. The resulting network structure reflects natural and infrastructural features as well as taking account vital biotope networks and watercourses (Fig. 4.27). Environmental functions of resulting subareas are spatially analyzed and connected with urban development targets. By doing so, a network consisting of functional spaces, corridors, nodes and green connectivity axes is defined, mirroring the existing features of the natural landscape as well as the polycentric organization of the city of Dresden (LH Dresden 2014). The aim is to provide ecosystem services throughout the entire city by preserving, developing and interlinking existing and new elements of the green infrastructure, which also includes green brownfields. The network, consisting of green space, is to secure the habitat network and a wide spectrum of environmental functions and ecosystem services (e. g. provision of fresh air, flood protection) that are important for urban ecology. At the same time, characteristic urban structures and landscapes as well as diverse recreation potentials for the residents are secured, and the provision of habitats for plants and animals will improve (LH Dresden 2014).

With this '*Leitbild*' (overall concept), the city of Dresden reacts on the one hand to challenges like shortage of resources, land consumption, demographic change, climate adaptation and mitigation, and the loss of biodiversity. On the other hand, a high environmental quality and high quality of life is aimed at by securing natural resources by means of keeping, enlarging and linking urban green spaces.

Such a strategic spatial development model can help define a concept of urban development and the future spatial structure of a city (Yu et al. 2011). At the same time, it assists the decision-making process on whether to preserve a particular brownfield site as a green space or to allow redevelopment. An example in Box 4.6 shows how conceptual design for including elements of green infrastructure can help to lower the temperature even in building areas.

The '*Leitbild*' (overall concept) "Dresden—the compact city in an ecological network" can be seen as a good practice example, as it meets the main green infrastructure principles addressing aspects of green structure (integration, connectivity, multi-scale and multi-object approach, multifunctionality) summarized by Hansen and Pauleit (2014). Thus, it can be regarded as a vital approach to achieve compact and green cities.

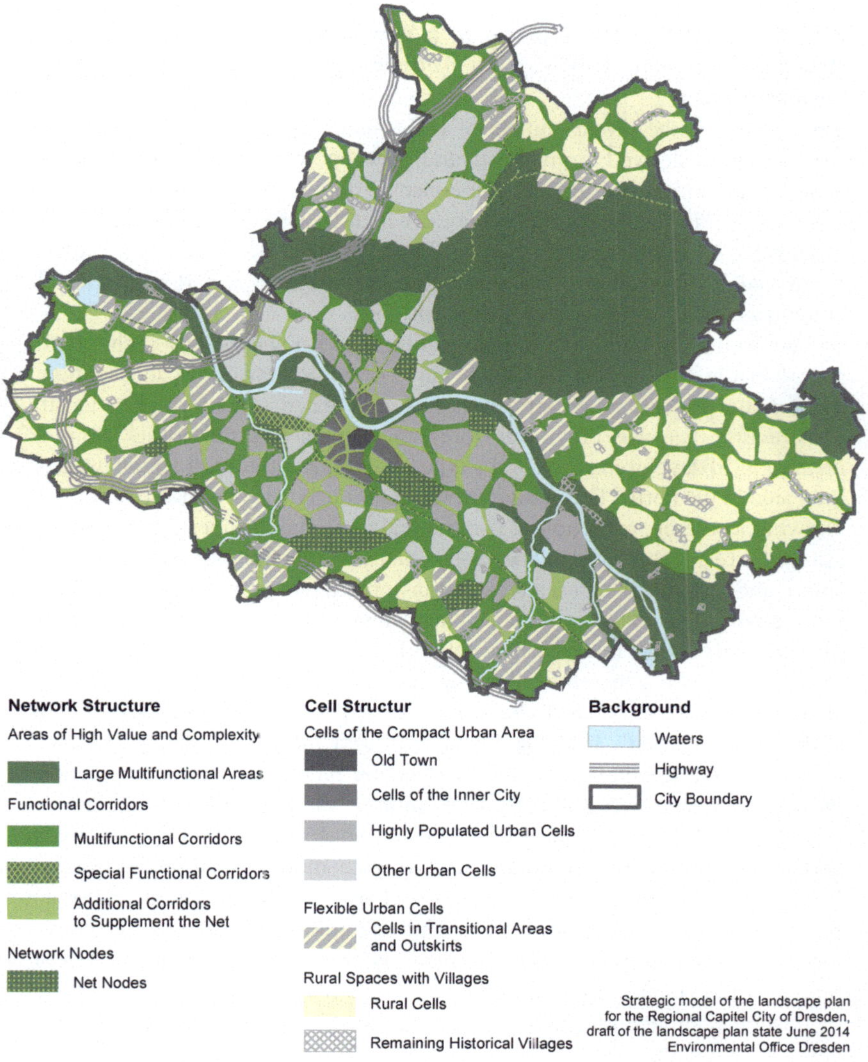

Fig. 4.27 The 'Leitbild' (overall concept) "Dresden: the compact city in an ecological network" of the draft Dresden landscape plan. Source: LH Dresden 2014, after LandschaftsArchitekt Paul

Box 4.6: Modeling the impacts of a new city quarter on a local construction plan using EnviMet (*by Wolfgang Wende, Adrian Hoppenstedt, Stephan Becsei*)

The impact on the urban climate of a German localsdevelopment and construction plan derived from the overall concept was simulated and tested using an urban climate model. The concept follows the spatial vision of the compact city in the ecological network, which underpins the landscape plan of Dresden's Environmental Agency. Wende et al. (2013) investigated how a green vision of urban development could be designed as part of a redevelopment of the downtown area 'Robotron Site'.

The draft local construction plan in Fig. 4.28 (a) shows how such a theoretical vision of a "compact city in an ecological network" can be implemented. Large-scale green infrastructures are created whilst a higher-density construction than was previously foreseen is planned in the north. The pink areas of infill development show the type of building usage; residential usage is indicated in the north. The orange-brown areas designate mixed-use buildings, allowing a certain proportion of commercial and small business in this district alongside residential space. The largest areas of this plot are zoned as green space, and thus must be unsealed and transformed into public green space. The individually depicted trees indicate sites where at least one new tree must be planted. This shows how the conceptual local development and construction plan foresees the generation of comprehensive green infrastructures, which should also serve to connect Dresden's city center areas with the neighboring park 'Großer Garten'. The plan's goals will be achieved by placing roadways below ground. Development and construction plans can include stipulations on plant characteristics and the preservation of valuable green structures.

Figure 4.28 (b) shows the microclimatic impacts of the plan in comparison to the initial situation using the microclimate model EnviMet (Bruse and Fleer 1998). Obviously, the climatic effect of the plan can be an important factor for climate change adaptation. This example shows how landscape and urban development planning in Germany is able to make a valuable contribution to the creation of green cities.

Several architectural ideas can reduce the climatic vulnerability of such new city quarters:

- Green walkways covered by twiners for pedestrian well-being (Fig. 4.29 left)
- Vegetated building envelopes (roof, façade, balcony) conditioning the interior climate
- Collection of rainwater and its usage for cooling and greening
- Movable façade covers allowing shading or irradiation by intelligent control
- Groundwater management (possible infiltration) and soil protection in open spaces

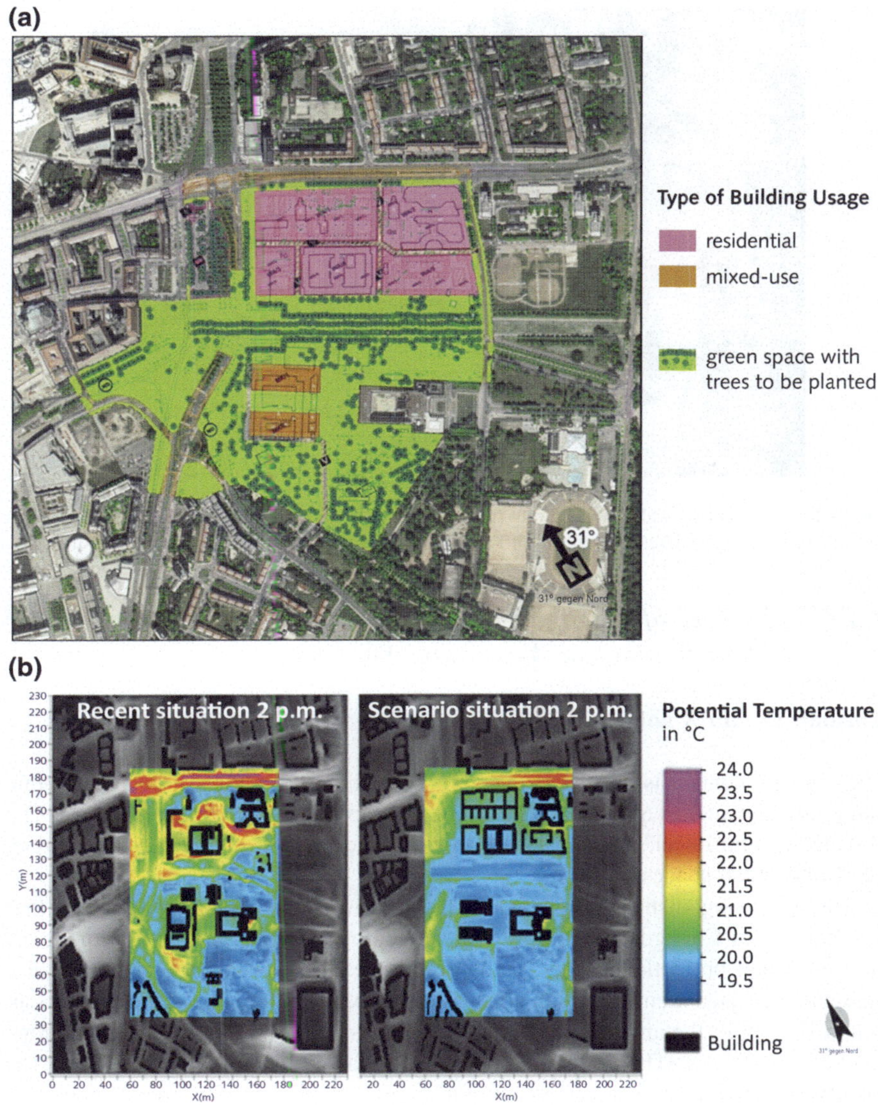

Fig. 4.28 **a** Green City Lab Dresden—conceptual design (scenario) for the development of a green hub and for generating green infrastructure in the ecological network Dresden. **b** Comparison of modeled midday air temperatures (at height 1.2 m) on a hot summer day. Actual state (*Left*) versus scenario (*middle-right*). *Source* Wende et al. 2013

Fig. 4.29 *Left* Green walkway in Frankfurt reducing the noonday heat by twiners © Stephan Becsei; *right*: alley in Dresden with trees shadowing road and walkway. © Robert Bendne

4.3.7 The City of Chengdu and the Wenjiang District: Urban Ecological Assets and Services

Xiushan Li and Xiaoliang Shi

The city of Chengdu is located in the southwestern part of China. It is an economic and political center of Southwest China and the capital of the Sichuan province. The climate is subtropical, and thus there is rich green infrastructure, such as rivers, lakes and green spaces.

The Wenjiang district is located in the northwest of Chengdu. It is the first satellite city of Chengdu, the first provincial ecological zone and the first model of environmental protection in the Sichuan province. It is the major component of a national ecological demonstration zone (Fig. 4.30). The Wenjiang district covers an area of 277 km^2, the annual average temperature is 16.9 °C, and the annual average rainfall is 740 mm. There are four rivers in the area, and the vegetation consists mainly of artificially designed vegetation types. At present, the green vegetation coverage rate is 41.26%, there are public green areas of 410 ha in the Wenjiang district, there are more than 1300 species of flowers and trees, and a plant area of 9860 ha. Chengdu is famous for its flowers and trees. The green space, rivers, wetlands, nursery stock flowers, orchards, the bamboo house Lin Pan and other kinds of "blue-green infrastructure" constitute a unique urban ecological system in the Wenjiang district.

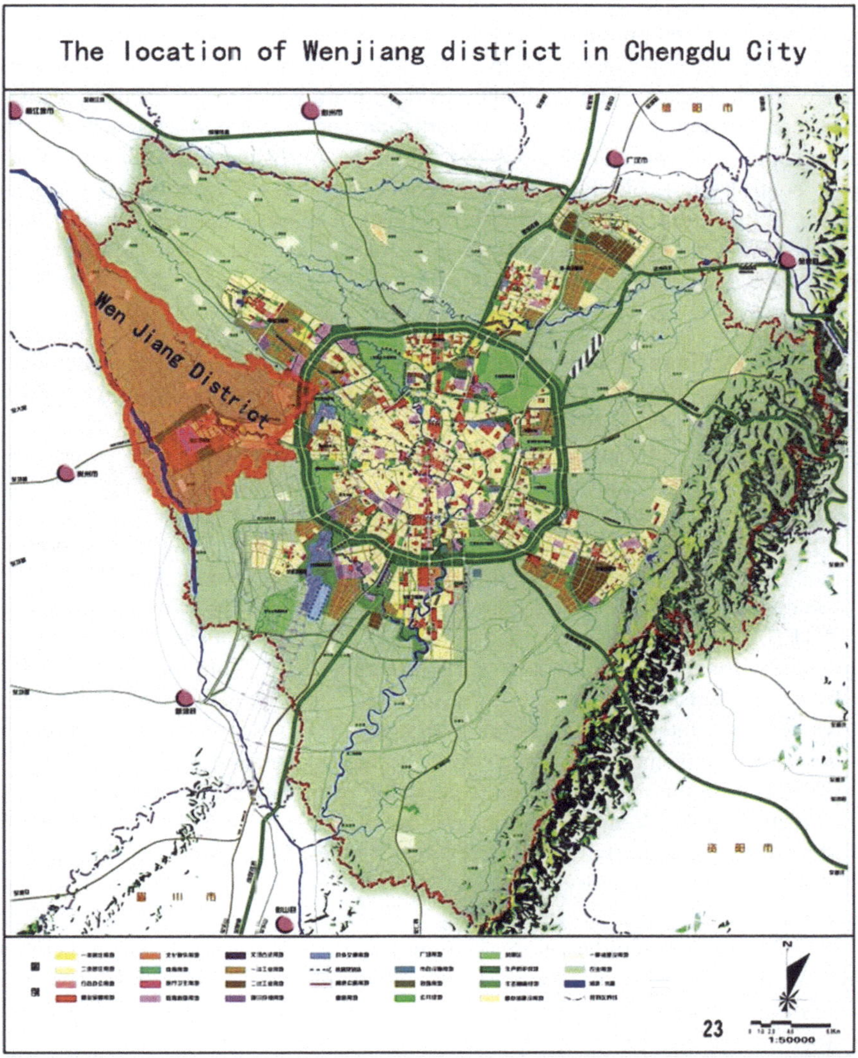

Fig. 4.30 The Wenjiang district, located in the northwest of Chengdu's city area. Source: City of Chengdu, modified

Ecosystem Service Value in the Wenjiang District

In order to increase urban ecosystem functions and services in the Wenjiang district, the local government authorized the Chinese Research Academy of Environmental Science (CRAES) to assess the urban ecosystem services in the Wenjiang district. The economic value of the ecosystem for human well-being has been analyzed and evaluated. The index system of ecosystem services that was thus developed contains three aspects of function and 14 indicators. The supply services mainly includes agricultural products, forest products, livestock products, fishery products, water resources, renewable energy, and seven other indicators (e.g., flowers, nursery stock, bonsai). The regulating services mainly includes indicators, such as water conservation (including water regulation and water purification, etc.), soil conservation, air purification (including fluoride, nitrogen oxide absorption and noise control, etc.), biodiversity maintenance, carbon sequestration and oxygen release, and disaster control and prevention (plant disease and pest control, flood control, etc.). The cultural services include recreational value and landscape aesthetic value (Tables 4.7 and 4.8).

Box 4.7 Willingness to pay for recreational ecosystem services in urban areas: a case study in the Wenjiang district of Chengdu, China (cf. Cao et al. 2016; Li et al. 2016)
The residents' willingness to pay (WTP) for the recreation services of urban ecosystems and its influencing factors were analyzed using the Logit model and the Oprobit model, and then the economic value of recreation services was calculated by the contingent valuation method (CVM). The results show that: (1) The average probability of respondents expressing a WTP in Chengdu and the Wenjiang district were 60.3 and 69.1% with an average WTP of 127.1 Yuan and 142.5 Yuan per year per person, respectively. (2) The probability and amount of the residents' WTP is not only influenced by respondents' income level, but also reflects the subjective evaluation by residents of internal infrastructure services in urban ecosystems and the degree of convenience, as well as the residents' demand for the services. (3) Based on the maximum WTP, the value of urban ecosystem services in recreation in the Wenjiang district was about 410 million Yuan. Finally, we suggest to integrate the urban recreational value of ecosystem services into eco-civilization/green city planning.

In 2015, we assessed the value of biodiversity maintenance and the recreational value of urban ecosystem services in the Wenjiang district (Box 4.7). The total WTP value for biodiversity maintenance is 9.1895×10^8 Yuan (Li et al. 2016), and the total recreational value is 4.1×10^8 Yuan (Cao et al. 2016).

Implementation of Ecosystem Services in the Wenjiang District
The urban ecosystem services evaluation showed that green space has a high value in the Wenjiang district. We suggest that the urban ecosystem services evaluation should be incorporated into the urban functional area division and industrial development planning during the implementation of ecological civilization. The ecological function area is divided into an eco-tourism zone, the Straits Science and Technology Park, a modern service industry park, and a reserved park for industrial upgrading and technological innovation. The northern part of the Wenjiang district belongs to the ecological manufacturing zone, including modern agriculture, flower

Table 4.7 Index system of accounting for ecosystem services

Serial number	Functional category	Accounting program	Instructions
1	Supply function	Agricultural products	Primary products of agricultural system, such as corn, rice, rapeseed and other grain; beans, potato; oil; cotton; hemp; sugar; tobacco leaf; tea; medicinal materials; vegetables; fruit, etc.
2		Forest products	Forest products, such as wood, rubber, raw lacquer, tea seed, etc.
3		Animal husbandry products	Grazing, in captivity, or a combination of these two ways, raising livestock and poultry for animal products, such as cattle, horses, donkeys, swine; dairy; eggs; honey; cocoons, etc.
4		Fishery products	Fish, shrimp, crab, shellfish, algae, etc.
5		Water resources	Freshwater resources, such as agricultural water and domestic water, industrial water and ecological water use, etc.
6		Ecological services and energy provision	Refers to the biological species and ecosystems; contains renewable energy, as well as the quality of oceans, lakes, rivers
7		Other services	Flowers, nursery stock, bonsai, etc., the value of these resources is usually set according to the demand of beneficiary
8	Regulating function	Flood water regulation	Provided by ecosystems through the process of structure and rainfall interception, savings, enhanced soil infiltration, effective conservation of soil water and groundwater complement, adjustment of the flow of rivers
9		Soil conservation	Ecosystem service through processes to reduce the rain erosion ability, reduce soil erosion
10		Purification of the atmosphere	Ecosystem service through absorption, filtration and decomposition of pollutants in the atmosphere and provision of oxygen, e.g., forests provide fluoride anions, absorb sulfur dioxide and nitrogen oxides and reduce noise and dust
11		Biodiversity maintenance	Birds, mammals, plants and insect species maintenance
12		Carbon sequestration, oxygen release	Forest ecosystems can fix carbon and release oxygen
13		Control of plant diseases and insect pests	Ecosystem service to increase the number of natural enemies, reduce locusts and achieve the ecological effect of pest control through improving the level of species diversity
14	Cultural function	Natural landscape	Value of human recreation, landscape aesthetics, inspiration, and education

Table 4.8 Ecosystem service value in the Wenjiang district in 2013

Functional category	Ecosystem service	Specific values	Value (10^8 yuan)	Total value (10^8 yuan)	Proportion (%)
Supply service function	Supply service	Agricultural products	0.845	122.111	24.98
		Forest products	0.028		
		Animal husbandry products	3.243		
		Fishery products	0.278		
		Water resources	7.416		
		Ecological services and energy provision	0.069		
		Other products (e.g., flowers, nursery stock)	110.232		
Regulating function	Water conservation	Value of water storage	0.7353	0.878	1.80
		Value of purifying water quality	0.1425		
	Soil conservation	Value of topsoil	0.010	1.040	2.28
		Value of saved fertilizer	1.030		
	Purification of the atmosphere	Value of negative ions	0.002	8.912	18.23
		Value of absorption of atmospheric pollutants and dust	8.150		
		Value of noise reduction	0.760		
	Carbon sequestration	Value of carbon sequestration	27.720	27.720	56.71
	Oxygen release	Value of releasing oxygen	0.010	0.010	0.02
	Plant disease and insect pest control	Value of plant disease and insect pest control	0.020	0.020	0.04
Total				48.8831	100

and tree cultivation, and it is designed to fully exploiting the recreation and recreation potential of the green infrastructure of forest, wetland and greenway. This attracts visitors from Chengdu to the leisure and recreation area while promoting local economic development.

The modern service industry park is the seat of the government and a residential, commercial and cultural area, whose main development direction is to develop and improve the service industry's level and quality, improve the living environment and improve residents' Happiness Index.

The Straits Science and Technology Park is a modern industrial park, intended to improve the industrial structure and develop low-carbon/low-pollution and high value-added modern industrial enterprises. The Science and Technology Park's main aim is to attract industrial upgrading and promote research and development.

4.3.8 Bonn: Intercommunal Project 'Green C' and the Integrated Action Plan 'Green Infrastructure'

Jonas Michels, David Baier and Ralf-Uwe Syrbe

On a total area of 4415 km^2, the population of the Köln-Bonn region amounted to 3,573,500 in 2012 (Region Köln-Bonn 2012) and is growing, particularly along the Rhine axis as the main traffic and natural backbone. The extension of the urban area increases the value of remaining open space in the proximity of residential areas. Open spaces are getting rare and have to be used multifunctionally. The urban agglomeration hosts many people looking for rest and recreation. The run on the recreational areas, but also on the natural reserves and rural landscape is significant, particularly on weekends and holidays. These open spaces are important:

- to ensure enough regional food for the population and thus to protect urban agriculture.
- as nearby public space for the city quarters with increasing residential (and thus building) density to enable functional local recreation for demographic groups not yet taken into consideration, such as the youth.
- for nature and species protection, including the necessary environmental education as a basis for sustainable development.

This section presents an attempt at a solution of the dilemma arising between shortage of space and the need for preserving green in a growing city. 'Green C' is the name of a collaboration project between the city of Bonn and some neighboring municipalities to create a continuous system of biotopes and recreational sites for the inhabitants that is to enhance quality of life and advance the soft local factors of the region. The attractiveness of scenic quality (Fig. 4.31) is an increasingly important criterion for the decision on moving or staying in a region and regarding it as home. The preservation of common recreational possibilities, nature, and farming is essential for the quality of green infrastructure. Coordinated development of open space beyond the municipal edges seems to be useful so as to complement communal urban planning.

The project 'Green C' was the first successful initiative in the 'Regional 2010' framework that picked up landscape planning issues and aimed at the protection of open space on a voluntary basis agreed between several municipalities of the Rhein-Sieg county (Alfter, Bornheim, Troisdorf, Sankt Augustin and Niederkassel)

Fig. 4.31 Redesign of the ferry landing Mondorf as an essential project part "Bridging the Rhine". Source: City of Bonn

and the city of Bonn. A key issue was the retention of non-urbanized green spaces despite the competition for investments and settlements in this dynamically developing region. These green spaces have been connected to improve alternative traffic as well as for the movement of animals. The entire green system surrounds the most densely settled areas of Bonn in the form of a semicircle (hence the name 'Green C') consisting of the element types links, stations, gates, and edges as shown in the map section (Fig. 4.32).

Starting from the 'Bridging the Rhine' ferry near Mondorf (Fig. 4.31, crossing the river band in the map) a continuous system of walking paths and cycleways extends into the region on both sides of the river with several further branchings. After completion, the system will connect the nature park 'Rheinland' in the west with the nature park 'Siebengebirge' in the southeast of the planning area (dark green in the map) via the nature protection area 'Siegaue' (Fig. 4.32).

Link: getting connected

The so-called 'Link' is the connecting pathway of the 'Green C' system. The pathway features different conditions and invites visitors to a guided nature discovery by walking, cycling or inline skating. It largely uses existing paths that were merely improved or renewed. Only some small gaps had to be closed in the system in order to cross highways, to access open spaces and to guide users on safe ways.

Fig. 4.32 Interactive Map of the 'Green C' with elements explained in the text. Source: City of Troisdorf http://www.gruenes-c.de/karte/

Stations

People get interesting information about nature and landscape at the so-called 'Stations' from information boards along the Link. The boards present and explain history, landscape change and recent land use along with their consequences for biological diversity, as well as the peculiarities of flora and fauna.

Gates and Edges

'Gates' enable the access to the Green C system and signify the junction between settlement and open landscape. The 'Edges' are an approach to maintenance of open space. Edges are the interface from urbanized area to open landscape. Their features should be as natural as possible. Depending on the local situation, they can vary in width and even partly fulfill functions such as those of recreation areas themselves. Footpaths along the settled Edges invite visitors to a take a walk in the evening or to spend some time. This is a way to protect coherent agricultural areas and refuges for animals and the rare flora of open space.

After implementation of 'Green C', an integrated action plan 'Green Infrastructure' is being developed, based on the most successful parts of the initial project. Its purpose is to develop the open space within its main part, depending on the demand of the neighborhoods concerned. The main objective is to maintain and sustainably shape the open space as an attractive location factor for a growing population. The 'Green C' is the backbone of the Green Infrastructure under development in the region and should be strengthened. Main upcoming topics of the integrated action plan 'Green Infrastructure' are urban agriculture, local recreation, and nature protection. Stakeholders and common people from different municipalities identify, discuss and select objectives and measures in decentralized workshops. The example shows that maintenance and development of green spaces does not end at the borders of municipalities. The voluntary intercommunal cooperation in the framework of the Green C points a way for the future progress of the region.

References

Barthel S, Parker J, Ernstson H (2013) Food and green space in cities: a resilience lens on gardens and urban environmental movements. J Urb Stud 52(7):1321–1338

Becker C, Giseke U, Herrmann A, Neuhaus A (2015) Konzeptgutachten Freiraum München 2030. Entschleunigung – Verdichtung – Umwandlung. Draft December. www.muenchen.de/rathaus/dms/Home/Stadtverwaltung/Referat-fuer-Stadtplanung-und-Bauordnung/Veranstaltungen/ZfS_2016/FRM2030_WEB.pdf. Accessed 15 June 2016

Beijing Municipal Statistical Bureau (2010) Beijing statistical yearbook. China Statistics Press, Beijing (in Chinese)

BfN—Bundesamt für Naturschutz (2008) Landscape planning—the basis of sustainable landscape development. Bundesamt für Naturschutz, Leipzig (Federal Agency for Nature Conservation). http://www.bfn.de/fileadmin/MDB/documents/themen/landschaftsplanung/landscape_planning_basis.pdf. Accessed 18 Aug 2016

BfN—Bundesamt für Naturschutz (2012) Daten zur Natur 2012. Bundesamt für Naturschutz (Federal Agency for Nature Conservation), Landwirtschaftsverlag Münster

BfN—Bundesamt für Naturschutz (2015) Naturschutz und Landschaftspflege in der integrierten Stadtentwicklung. Argumente, Positionen, Hintergründe. Bundesamt für Naturschutz (Federal Agency for Nature Conservation), Bonn

Bläser K, Danielzyk R, Fox-Kämper R, Funke L, Rawak M, Sondermann M (2012) Urbanes Grün in der integrierten Stadtentwicklung. Strategien, Projekte, Instrumente. the North Rhine-Westphalia Ministry for Economic Affairs, Energy, Building, Housing and Transport, Düsseldorf

Böhm J, Böhme C, Bunzel A, Kühnau C, Reinke M (2016) Urbanes Grün in der doppelten Innenentwicklung. BfN-Skript 444, Bonn, Bad Godesberg

Breuste J (2016) Was sind die Besonderheiten des Lebensraumes Stadt und wie gehen wir mit Stadtnatur um? In: Breuste J, Pauleit S, Haase D, Sauerwein M (eds) Stadtökosysteme. Funktion, Management Entwicklung. Springer, Berlin, pp 85–128

Breuste JH, Artmann M (2015) Allotment gardens contribute to urban ecosystem service: case study, Salzburg, Austria. J Urb Plan Dev 141(3):A2015001

Breuste J, Pauleit S, Haase D, Sauerwein M (2016) Stadtökosysteme. Funktion, Management, Entwicklung, Berlin

Bruse M, Fleer H (1998) Simulating surface-plant-air interactions inside urban environments with a three dimensional numerical model. Environmental Modelling & Software, (13):373–384

BZ Berlin (2016) Chillen wie die Spießer. Berlins Laubenpieper werden immer jünger und hipper. http://www.bz-berlin.de/stadtleben/sommer-stadtleben/berlins-laubenpieper-werden-immer-juenger-und-hipper. Accessed 21 June 2016

Cao XL, Liu GH, Zhang Y, Li XS (2016) Willingness-to-pay for recreation ecosystem services in urban areas: a case study in Wenjiang District of Chengdu City, China. Acta, Ecologica, Sinica. In press (in Chinese)

Castle H (2008) Dongtan, China's flagship eco-city: an interview with Peter Head of Arup. Archit Des 78(5):64–69
Chang J, Feng S (2008) Strategies on redevelopment of mining city industrial wasteland. Urb Dev Stud 02(2008):54–57
Chang J, Koetter T (2005) From abandoned mine land to landscape park. J China Coal Soc 30:399–402
Chang J, Wende W, Luo PJ, Deng YY (2011) Re-use of the mining wasteland. Tongji University Press, Shanghai
City of Munich (2012) Wohnen in München V. Wohnungsbauoffensive 2012–2016. Referat für Stadtplanung und Bauordnung, Munich
City of Munich (2015a) Bevölkerungsbestand. Monatszahlen aus 2015. www.muenchen.de/rathaus/dms/Home/Stadtinfos/Statistik/bevoelkerung/Monatszahlen/bev_stand_05.pdf. Accessed 24 July 2015
City of Munich (2015b) Demografiebericht München – Teil 1. Analyse und Bevölkerungsprognose 2013 bis 2030. Referat für Stadtplanung und Bauordnung, Munich
City of Munich (2015c) 40 Jahre Wettbewerb Mehr Grün für München. https://www.muenchen.de/rathaus/Stadtverwaltung/baureferat/freizeit-sport-natur/wettbewerb-mehr-gruen/preistraeger-2014-2015.html. Accessed 15 June 2016
City of Munich (2016) Schutzgebiete. www.muenchen.de/rathaus/Stadtverwaltung/Referat-fuer-Stadtplanung-und-Bauordnung/Natur-Landschafts-Baumschutz/Schutzgebiete.html. Accessed 28 Jan 2016
Davies C, Hansen R, Rall E, Pauleit S, Lafortezza R, De Bellis Y, Santos A, Tosics I (2015) Green infrastructure planning and implementation. The status of European green space planning and implementation based on an analysis of selected European city-regions. http://greensurge.eu/working-packages/wp5/files/Green_Infrastructure_Planning_and_Implementation.pdf. Accessed 02 Sept 2016
De Roo G (2000) Environmental conflicts in compact cities: complexity, decision making, and policy approaches. Environ Plann B: Plann Des 27:151–162
Deng X, Peng XC, Tan CM (2010) The review of the functions and characteristics of roof greening, and its current situations and problems in China. Acta Scientiarum Naturalium Universitatis Sunyatseni s1:99–101
Drescher A (2001) The German allotment gardens—a model for poverty alleviation and food security in southern African Cities? In: Proceedings of the sub-regional expert meeting on urban horticulture, Stellenbosch, South Africa, FAO, University of Stellenbosch, 15–19 Jan 2001
Droste N, Schröter-Schlaack C, Hansjürgens B, Zimmermann H (in print) Implementing nature-based solutions in urban areas: financing and governance aspects. In: Kabisch N, Bonn A, Korn H, Stadler J (eds) Nature-based solutions to climate change in urban areas. Springer Publisher, Berlin
EC—European Commission (2012) Guidelines on best practice to limit, mitigate or compensate soil sealing (Commission staff working document). EC, Brussels
EEA—European Environment Agency (2011) Green infrastructure and territorial cohesion. The concept of green infrastructure and its integration into policies using monitoring systems. EEA Technical Report 18, Copenhagen
Espey M, Owusu-Edusei K (2001) Neighborhood parks and residential poperty values in Greenville, South Carolina. J Agric Appl Econ 33(3):487–492
Feng SS, Chang J, Hou W (2016) A framework for setting restoration priorities for coal subsidence areas based on green infrastructure (GI). Acta Ecol Sin 26:2724–2731
Gärtner S (2015) Towards BiodiverCity Mainz—current processes in the capital of Rheinland-Palatinate. Presentation to the 8th Sino-German workshop: integration of ecological aspects in city planning, Berlin, June 2015
Gawel E (1995) Die kommunalen Gebühren. Duncker & Humboldt, Berlin

Gawel E (2016) Environmental and resource costs under Article 9 Water Framework Directive. Challenges for the implementation of the principle of cost recovery for water services. Duncker & Humblot, Berlin

Georgi JN, Dimitriou D (2010) The contribution of urban green spaces to the improvement of environment in cities: case study of Chania, Greece. Build Environ 45:1401–1414

Georgieva Y (2015) 10 Cities that are reinventing the relationship with their rivers. Posted by LAN 25 Aug 2015 in environment posts, landscape architecture posts, Riverside. https://landarchs.com/10-cities-that-are-reinventing-the-relationship-with-their-rivers/

Geyler S, Bedtke N, Gawel E (2014) Sustainable rainwater management in existing settlements. GWF Wasser Abwasser 155(1):96–102, 155(2):214–222

Hansen R, Pauleit S (2014) From multifunctionality to multiple ecosystem services? A conceptual framework for multifunctionality in green infrastructure planning for urban areas. Ambio 43 (4):516–529

Heiland S, Rößler S, Wende W (2016) 9.4.1 Formelle planerische Instrumente. In: Kowarik I, Bartz R, Brenck M (eds) Ökosystemleistungen in der Stadt: Gesundheit schützen und Lebensqualität erhöhen. Technische Universität Berlin, Helmholtz-Zentrum für Umweltforschung - UFZ, Berlin, Leipzig, pp 236–245

Hu T, Cheng D, Xiong H, Qin Q (2016a) Construction of Xuzhou elderly public service facilities under the perspective of age-friendly. J Jiangsu Normal Univ (Nat Sci Ed) 01(2016):15–18

Hu T, Shen S (2014) Research progress and construction practice of age-friendly city. Mod Urb Res 09(2014):14–20

Hu T, Shen S, Chang J (2016b) Research progress and construction practice of age-friendly city. Case studies of New York City, the U.S. and London, Canada. Urb Plan Intl 04(2016):127–130

Ignatieva MA, Xiu N, Yang F (2015) Chinese Urban Green Areas: Classic Gardens to a Globalized Landscape. The Nature of Cities, April 5, 2015

Ioannou B, Morán N, Sondermann M, Certomà C, Hardman M (2016) Grassroots gardening movements: towards cooperative forms of green urban development? In: Bell S, Fox-Kämper R, Keshavarz N, Benson M, Caputo S, Noori S, Voigt A (eds) Urban allotment gardens in Europe. Routledge, New York, pp 62–90

Kambites C, Owen S (2006) Renewed prospects for green infrastructure planning in the UK. Plan Pract Res 21:483–496

Keeley M, Koburger A, Dolowitz DP, Medearis D, Nickel D, Shuster W (2013) Perspectives on the use of green infrastructure for stormwater management in Cleveland and Milwaukee. Environ Manage 51:1093–1108

Keshavaraz N, Bell S (2016) A history of urban gardens in Europe. In: Bell S, Fox-Kämper R, Keshavarz N, Benson M, Caputo S, Noori S, Voigt A (eds) Urban allotment gardens in Europe. Routledge, New York, pp 8–32

Küchler J, Shao Y (2017) Vom Volkspark zur Stadtlandschaft. Ein Jahrhundert modernes Stadtgrün in China. Stadt + Grün 2:35–41

Leung KH (2005) Wherever there is a road, there is greening. Hong Kong Highways Department Newsletter Landscape vol 2, pp 1–4. http://www.hyd.gov.hk/ENG/PUBLIC/PUBLICATIONS/newsletter/2005/Issue2/E205A10.pdf. Accessed 15 Apr 2015

LGSMP—Leading Group of the Shanghai Master Plan (2015) Outline of the Shanghai Master Plan (2015–2040). Shanghai. https://www.img.thupdi.com/news/2016/01/1453791519864576.pdf. Accessed 20 Aug 2016

LH Dresden—Landeshauptstadt Dresden (2014) Landschaftsplan der Landeshauptstadt Dresden. Draft (as of June 2014). http://www.dresden.de/de/stadtraum/umwelt/umwelt/landschaftsplan/unterlagen_landschaftsplan.php. https://www.dresden.de/media/pdf/umwelt/LP_Erlaeuterungstext.pdf. Accessed 2 Sept 2016

LH Dresden—Landeshauptstadt Dresden (2016a) Zukunft Dresden 2025+. Integriertes Stadtentwicklungskonzept Dresden (INSEK). http://www.dresden.de/media/pdf/stadtplanung/stadtplanung/spa_insek_Broschuere_DD_2025_final_Internet_n.pdf. Accessed 2 Sept 2016

LH Dresden—Landeshauptstadt Dresden (2016b) Kommunale Statistikstelle. https://www.dresden.de/de/leben/stadtportrait/statistik/. Accessed 6 June 2017

Li J, Wang Y, Song Y-C (2008) Landscape corridors in Shanghai and their importance in urban forest planning. In: Carreiro MM, Song Y-C, Wu J (eds) Ecology, planning, and management of urban forests: international perspectives. Springer Series on Environmental Management, New York, pp 219–239. https://www.faculty.ecnu.edu.cn/…/613b11a9-674e-4766-978d-aa25da954b81.pdf.x. Accessed 18 Aug 2016

Li L (2010) Cost control of the conservation and management for urban green spaces. J Jiangsu For Sci Technol 1:32–35

Liu W, Huang X (2007) Shanghai urban planning. Shanghai

Li XS, Liu GH, Cao XL, Du LS (2016) Factors influencing willingness to pay for the urban biodiversity and its value evaluation: Wenjiang District in Chengdu City as an example. Environ Prot 44(03–04):18–25 (in Chinese)

Lin Z, Chang J, Wang W (2009) Research on planning of subsidence land's ecological restoration in Xuzhou Mining Area under the framework of urban-rural coordination. Mod Urb Res 10 (2009):91–95

Mathey J, Rößler S, Banse J, Lehmann I, Bräuer A (2015) Brownfields as an element of green infrastructure for implementing ecosystem services into urban areas. J Urb Plan Dev 141(3): A4015001-1 to A4015001-13. doi:10.1061/(ASCE)UP.1943-5444.0000275

Masterplan of New Chongming City (2003)

Ministry of Construction (2002) Standard for classification of urban green space, CJJ/T 85-2002, J185-2002. Ministry of Construction Press, Beijing (in Chinese)

Mu FY, Zhang ZX, Chi YB, Liu B, Zhou QB, Wang CY et al (2007) Dynamic monitoring of built-up area in Beijing during 1973–2005 based on multi-original remote sensed images. J Remote Sens 11(2):257–268 (in Chinese)

Municipal Sewage Works of the City of Munich (2012) Niederschlagswassergebühren. http://www.muenchen.de/rathaus/dam/jcr:a740b22a-74cd-4093-ba58-a17fed629cdc/niederschlagswassergebuehren.pdf. Accessed 18 Jan 2017

National Bureau of Statistics of China, Department of Urban & Social Economic Survey (2009) China city statistical yearbook. China Statistics Press, Beijing

Pawlikowska-Piechotka A (2011) Child-friendly urban environment and playgrounds in Warsaw. Open House Int 36(4):98–110

Qu AX, Bian ZF, Zhu CG et al (2009) Study on the process and pattern of land use change in Xuzhou urban area. Geogr Res 1:97–108, 276

Region Köln-Bonn (2012) Die Region in Zahlen. http://www.region-koeln-bonn.de/de/region/zahlen-daten-fakten/index.html. Accessed 25 Jan 2017

REGKLAM—Regional Climate Change Adaptation Programme Dresden Region (2012) Managing risks, seizing opportunities The Dresden region faces up to climate change. Abridged version, draft, Dresden. http://www.regklam.de/fileadmin/Daten_Redaktion/Publikationen/REGKLAM_ManagingRisks_abridged.pdf. Accessed 2 Sept 2016

Rittel K, Bredow L, Wanka ER, Hokema D, Schuppe G, Wilke T, Nowak D, Heiland S (2014) Green, natural, healthy: the potential of multifunctional urban spaces. Final Report. R&D Project FKZ 3511 82 800 Federal Agency for Nature Conservation, 107pp

Rößler S, Albrecht J (2015) Umsetzung freiraumplanerischer Klimaanpassungsmaßnahmen durch stadt- und umweltplanerische Instrumente. In: Knieling J, Müller B (eds) Klimaanpassung in der Stadt- und Regionalentwicklung – Ansätze, Instrumente, Maßnahmen und Beispiele. Klimawandel in Regionen zukunftsfähig gestalten. oekom Verlag, Munich, Vol 7:243–270

Rößler S, Sondermann M, Herbst T (2016) 9.4.2 Informelle planerische Konzepte und Instrumente. In: Kowarik I, Bartz R, Brenck M (eds) Ökosystemleistungen in der Stadt: Gesundheit schützen und Lebensqualität erhöhen. Technische Universität Berlin, Helmholtz-Zentrum für Umweltforschung – UFZ, Berlin, Leipzig, pp 245–251

Rosol M (2012) Community volunteering as neoliberal strategy? Green space production in Berlin. Antipode 44(1):239–257

Rüger J, Gawel E, Kern K (2015) Reforming the German rain water charge—Approaches for an incentive-oriented but still workable design of the charge. GWF—Wasser Abwasser 156 (3):364–372

Santos R, Schröter-Schlaack C, Antunes P, Ring I, Clemente P (2015) Reviewing the role of habitat banking and tradable development rights in the conservation policy mix. Environ Conserv 42:294–305

SBA Design (2003) Masterplan of new Chongming city, Shanghai. http://www.sba-int.com/. Accessed 31 Dec 2013

Schröder A, Arndt T, Mayer F (2016) Naturschutz in der Stadt – Grundlagen, Ziele und Perspektiven. Nat Landschaft 7:306–313

Schröter-Schlaack C (2013) Steuerung der Flächeninanspruchnahme durch Planung und handelbare Flächenausweisungsrechte. Helmholtz-Zentrum für Umweltforschung—UFZ, Leipzig. Senate Department for Health, the Environment and Consumer Protection Kleingärten. Daten und Fakten. http://www.stadtentwicklung.berlin.de/umwelt/stadtgruen/kleingaerten/de/daten_fakten/index.shtml. Accessed 21 June 2016

Senatsverwaltung für Stadtentwicklung (2004) Kleingartenentwicklungsplan Berlin. Senate's Department of Urban Development http://www.stadtentwicklung.berlin.de/umwelt/stadtgruen/kleingaerten/downloads/KEP-TEXT_2004.pdf. Accessed 21 June 2016

Senatsverwaltung für Stadtentwicklung und Umwelt (no year) Kleingärten. Daten und Fakten. http://www.stadtentwicklung.berlin.de/umwelt/stadtgruen/kleingaerten/de/daten_fakten/index.shtml. Accessed 21 June 2016

Shanghai Municipal Statistics Bureau (SMSB) (2006) Shanghai Statistics. http://www.stats-sh.gov.cn/2004shtj/tjnj/tjnj2007e.htm. Accessed 2 April 2009

Shanghai Municipal Statistics Bureau (2014) Shanghai Statistical Yearbook 2014. http://www.stats-sh.gov.cn/data/toTjnj.xhtml?y=2014e. Accessed 8 May 2015

Shen S, Lin L, Jiang G (2012) Theoretical and empirical urban planning assessment. Southeast University Press, Nanjing

SMSB—Shanghai Municipal Statistics Bureau (2006) Shanghai Statistics. http://www.stats-sh.gov.cn/2004shtj/tjnj/tjnj2007e.htm. Accessed 2 Apr 2009

Sondermann M, Rößler S (2016) Wege zur Umsetzung – Integration von Ökosystemleistungen in Entscheidungen der Stadtentwicklung: 9.1 Leitbilder erstellen, Orientierung bieten. In: Kowarik I, Bartz R, Brenck M (eds) Ökosystemleistungen in der Stadt: Gesundheit schützen und Lebensqualität erhöhen. UFZ, Leipzig, pp 218–221. http://www.naturkapital-teeb.de/fileadmin/Downloads/Projekteigene_Publikationen/TEEB_Broschueren/TEEB_DE_Stadtbericht_Langfassung.pdf. Accessed 02 Sept 2016

SUPAB—Shanghai Urban Planning Administration Bureau, SUPDRI—Shanghai Urban Planning and Design Research Institute (no year) Summary of the Comprehensive Plan of Shanghai (1999–2020). Shanghai

Wang XX (2008) Beijing practices tackling climate change. Intl J Urb Sci 12(1):40–48

Wende W, Huelsmann W, Marty M, Penn-Bressel G, Bobylev N (2010) Climate protection and compact urban structures in spatial planning and local construction plans in Germany. Land Use Policy 27(3):864–868

Wende W, Rößler S, Held F (2013) Green City Lab. Leistungen von Stadtbäumen und -vegetation für einen stadtklimatischen Ausgleich - eine Modellierung am Beispiel Dresdens. In: Roloff A, Thiel D, Weiss H (eds) Aktuelle Fragen der Stadtbaumplanung, -pflege und -verwendung. Forstwissenschaftliche Beiträge Tharandt. Contributions to Forest Sciences, Beiheft 14

Wessolek G, Nehls T, Kluge B (2014) Bodenüberformung und Versiegelung. In: Blume H-P, Horn R, Thiele-Bruhn S (eds) Handbuch des Bodenschutzes. Wiley-VCH, Weinheim, pp 155–169

WHO—World Health Organization (1946) WHO definition of health. In: Preamble to the constitution of the world health organization as adopted by the International Health Conference, New York, 19–22 June 1946. Official Records of the World Health Organization 2, p 100

Wirth P, Cernic Mali B, Fischer W (2012) Post-mining regions in central Europe—problems, potentials, possibilities. In: Wujun L, Xiang H (eds) (2007) Shanghai urban planning. Shanghai. Oekom, Munich

Wirth P, Lintz G (2007) Strategies of rehabilitation and development in European mining regions. In: Good (Best) practice cases in regional development after mining and industry. Grazer Schriften für Geographie und Raumforschung 42:75–85

Wolch J, Jerrett M, Reynolds K, McConnell R, Chang R, Dahmann N et al (2011) Childhood obesity and proximity to urban parks and recreational resources: a longitudinal cohort study. Health Place 17(1):207–214

Xiao N (2015) Biodiversity and ecosystem services in Beijing City. 8th Sino-German workshop on biodiversity conservation, Berlin, 19 June 2015

Xie ZZ, Song CY (2014) Research on government intervention of ecological environment protection. J Changchun Univ Technol Soc Sci Ed 6:4–6

Xuzhou Statistics al Yearbook (2015) 2015 Xuzhou statistical bulletin of the national economy development. http://www.xz.gov.cn/zgxz/zwgk/20160316/008016002_cf4758ef-5b90-4c26-9de1-89df7bb30be1.htm. Accessed 16 Mar 2016

Yang J, McBride J, Zhou JX et al (2005) The urban forest in Beijing and its role in air pollution reduction. Urb For Urb Green 3:65–78

You W (2009) Shanghai urban greening. In: Land N (ed) (2009) Urbanatomy Shanghai 2009. Shanghai, pp 287–293

Yu D, Jiang Y, Kang M, Tian Y, Duan J (2011) Integrated urban land-use planning based on improving ecosystem service: Panyu case, in a typical development area in China. J Urb Plan Dev 137(4):448–458

Zhang B, Xie GD, Zhang CQ et al (2012) The economic benefits of rainwater-runoff reduction by urban green spaces: a case study in Beijing, China. J Environ Manage 100:65–71

Zhang B, Xie GD, Gao JX et al (2014) The cooling effect of urban green spaces as a contribution to energy-saving and emission-reduction: a case study in Beijing, China. Build Environ 76: 37–43

Zhang B, Xie GD, Li N et al (2015) Effect of urban green space changes on the role of rainwater runoff reduction in Beijing, China. Landscape Urb Plan 140:8–16

Zhang NN (2005) Maintenance problems and countermeasures of public green space in Xi'an urban area. Shanxi For Sci Technol 05:107–109

Zhou YH, Zhao J (2014) On the employment of economic tools in Chinese environmental management. Sci Econ Soc 2:62–64, 83

Zhu Y, Li L, Chen GH (2011) Roof greening policy incentives in China and abroad. Gardens 08:14–18

Chapter 5
Towards 'Green Cities'—Fields of Action and Recommendations

Karsten Grunewald, Tinghao Hu, Lennart Kümper-Schlake, Wei Hou and Qiaoqiao Xu

In order to secure and create urban green space, actions need to be coordinated across different policy areas and between institutions and persons involved in sustainable urban development as well as on different subnational planning levels (regional level, city-wide level, district/site level). The implementation needs to be seen as a joint task by authorities, citizens, enterprises, and other stakeholders. In this final chapter, recommendations are summarized on how to deal with the immense challenges and how to support stakeholders at work. These are based on the results of bilateral field work and the findings from the case studies. A guideline with regard to the integration of biodiversity and ecosystem services into municipal landscape planning, a brief discussion of the role of communication and cooperation between science, policy and practice as well as between countries such as China and Germany, and a set of research needs complete the final chapter.

K. Grunewald (✉)
Leibniz Institute of Ecological Urban and Regional Development, Dresden, Germany
e-mail: k.grunewald@ioer.de

T. Hu
China University of Mining and Technology, Xuzhou, China
e-mail: hth11211@163.com

L. Kümper-Schlake
Federal Agency for Nature Conservation (BfN), Bonn, Germany
e-mail: lennart.kuemper-schlake@bfn.de

W. Hou
Chinese Academy of Surveying and Mapping, Beijing, China
e-mail: houwei@casm.ac.cn

Q. Xu
East Asia Secretariat ICLEI, Seoul, South Korea
e-mail: qiaoqiao.xu@iclei.org

K. Grunewald et al. (eds.), *Towards Green Cities*, Cities and Nature,
DOI 10.1007/978-3-319-58223-8_5

5.1 How to Address the Challenges?

Karsten Grunewald, Martina Artmann, Olaf Bastian, Jürgen Breuste, Boping Chen, Tinghao Hu, Dennis Kalisch, Lennart Kümper-Schlake, Xiushan Li, Juliane Mathey, Stefanie Rößler, Ralf-Uwe Syrbe, Wei Hou, Henry Wüstemann and Suili Xiao

Urban green spaces contribute to sustainable development in manifold ways. Thus, they are essential prerequisites for livable, resident-, and future-oriented cities and municipalities. The concept of ecosystem services can be an integrative tool to help in planning, developing, and managing urban green spaces in compact cities (Artmann et al. 2017). It specifies particular social benefits provided to residents and local associations, thus linking beneficiaries with ecological assets (Fig. 5.1) stated in landscape planning (Sect. 5.2). But how can we integrate urban ecosystems into urban structure and form? Ecological processes need space and time. There is a great need for both ecological and economic development, but space in cities is limited.

For instance, the Shanghai urban green space system (Sect. 4.3.3), like many green areas in both Chinese and German cities, has not been fully evaluated with respect to its benefits for people. Urban green spaces and their management require substantial financial resources (e.g., planning, implementation, and management costs), and effective control is being demanded more and more, also from the political arena. As they will support ecosystem service balances, both, China and Germany, need to develop methods (and their implementation) for assessing the ecosystem services and biodiversity according to which social groups benefit from them and how many people really receive benefits.

Overarching challenges in developing greener cities in China and Germany were pointed out in Sect. 2.1. The recommendations for dealing with these

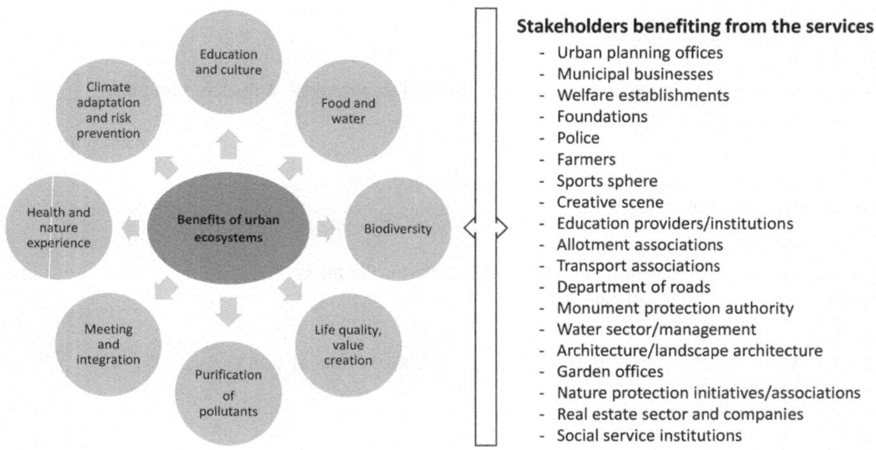

Fig. 5.1 Benefits provided by urban green spaces (*left*) and stakeholders benefiting from the services (*right*). © Karsten Grunewald

challenges are summarized in Box 5.1. Both countries are confronted with urbanization, but are very different in their current status, their dynamics and the forms, control and management of the urbanization processes. During phases of rapid urbanization, individual and societal welfare are boosting resource consumption, especially land and energy consumption. At the same time, both countries consider the urgent need to protect, develop, and enhance urban green spaces in quality and quantity in order to benefit from their ecosystem services as a means to increase human well-being in cities.

Box 5.1 Dealing with the challenges to develop greener cities in China and Germany (see Box 2.2, Chap. 2)

Find ways to

1. **Limit urban land consumption and change from quantitative to qualitative urban growth**

The limited resources must be efficiently used to achieve the aims of a good quality of life and a healthy environment for an increasing number of urban people. The effective use of all recourses consumed in cities, e.g., energy and matter, is an urgent need. Economic and physical urban growth can be organized to be based on existing urban structures and to minimize resource consumption. Green infrastructure, such as green belts, can support the limitation of urban expansion.

2. **Find an optimal urban form, integrating various kinds of nature**

A compact city with integrated urban green spaces is regarded as a sustainable urban form to reduce urban sprawl and the consumption of resources within the city, connecting it to an urban region.

3. **Create healthy urban living conditions**

With their location and ecosystem services, urban green spaces are essential parts of healthy urban living conditions. Germany has already reached high urban standards and tries to maintain and improve them. Healthy urban living conditions must be defined by standards, monitored, and implemented. China has much potential to overcome its deficits in this regard.

4. **Use urban green spaces and their management for adapting to climate change suitable structure and adjusted maintenance**

Urban green spaces can moderate negative thermal and hydrological effects, reduce effects of climate-related disasters, and can contribute to reducing health risks for vulnerable population groups (e.g., children, elderly, disabled people). Urban green spaces must be integrated into the neighborhoods, preferably in places with the highest need (highest negative climate effects, location of most vulnerable people).

5. **Ensure that all urban dwellers profit from benefits of urban green spaces**

New public urban green spaces should offer ecosystem services in demand to all urban dwellers, be accessibly located, extended, and improved where an urban majority of people can benefit (environmental justice).

6. **Engage people in planning process, decision-making, design, and management of urban green spaces**

Education and information sharing about urban green planning on green targets and benefits should be the first step of citizen involvement. Contributions from local people to neighborhood planning should always be welcome and taken seriously by the decision-makers, even when it makes the decision-making process slower or more complicated.

7. **Improve the usage and usability of existing scientific and practical knowledge for better decision-making and urban design**

The existing scientific and practical knowledge must be made available and accessible for practitioners. It should be integrated into standards that have to be fulfilled in design and decision-making.

8. **Reach green city targets with limited public budgets for greener cities**

New forms of self-sustaining green spaces and of public–private partnerships in establishing and managing them must be found. All forms of nature should be welcome. This requires education on urban nature benefits and risks. Public–private partnerships should be welcomed in managing urban green spaces and making them as accessible as possible. Citizens' protection and participation can be a valuable contribution to maintenance.

9. **Integrate all forms of nature into urban development for people's nature experiences and benefits**

Improved nature contact should consider not only the classical urban parks but also many other nature-like relics of pristine nature or rural landscape (agricultural land), horticulturally modified nature, and novel elements of new wild urban nature.

10. **Develop eco-districts and build new eco-cities**

Green-blue infrastructure is the core of the eco-city concept (Sect. 2.2) and makes up a significant part of the quality of urban life. Eco-cities satisfy demands of dwellers for urban environment, provide numerous ecosystem services such as supporting the adaption to climate change impacts, provide support for biodiversity, and are essential for environmental education and contact with nature. Eco-districts can be developed proceeding from the existing urban pattern or, mostly in China, newly designed optimal eco-cities can be built.

The governments of China and Germany are committed, as part of their areas of responsibility, to strengthening urban development through integrated and sustainable urban development (Sect. 2.2). In Germany, the 'Green Book' was presented in 2015, which, for the first time and across all departments, summarizes the current knowledge about urban green space (BMUB 2015). This was the beginning of a longer process, with which new, integrated strategies for urban green are developed and implemented. Based on this, a 'White Book' process was launched, which initiated a broad dialogue on recommendations for action toward green cities in Germany. The results were published in May 2017 (BMUB 2017b) and can be a guideline also for other countries such as China.

Overall visions and strategies might help to negotiate and focus on common principles and objectives of urban development among different stakeholders and sectors. Determined strategies and concepts addressing urban biodiversity and urban green spaces might bundle efforts of administrations and civil society to reach objectives and to implement measures appropriately. Implementing urban green spaces is a task of public duty of particularly high importance and should be appreciated accordingly. Therefore, one should use the power of terms like 'green infrastructure' and 'green city' in terms of planning, decision-making, negotiation, etc. Accordingly, *Leitbilder* (guiding visions for urban development) should be reframed, for instance, 'from a car-friendly city to an environmentally friendly city' or 'from urban sprawl to a compact city in a green network'. Further demands to be derived from the current 'Green Book/White Book' process in Germany (BMUB 2015; BUMB 2017b) and the main recommendations of the book in hand are:

- New urban development must be provided with ecological targets and means of monitoring and measuring of results should be guaranteed. Mainstreaming biodiversity and ecosystem services in urban development policies requires the establishment of adequate indicators and the integration of natural capital into national reporting and accounting systems. Based on case study findings, the following measures and targets are recommended: (a) setting threshold values for soil sealing in cities, (b) strict protection of existing high-quality urban green space (establishment of the 'no-net-loss principle' for urban green spaces, and (c) an increase of urban green spaces in both quantity and quality in order to ensure sufficient and equal access to green spaces.
- Green space development plans or 'master plans' for urban green as instruments for overall city planning are relevant, combining individual initiatives and strengthening integrative planning approaches. These must be aligned in the medium-term with a 'zero-hectare strategy' (balance new soil sealing with unsealing, no-net sealing) and 'no-net-loss strategy' (balance of ecosystem

services and biodiversity). Qian et al. (2016) consider the aspects of urbanization and land management systems that are unique to China and conclude that the current top-down directive and mandatory mode of control, which relies on the central government, has very limited effects. We think that a top-down, policy-led approach can be a proper way to ensure the optimum distribution of development and green space at the city level. But the local authorities need their own scope for action (Sects. 2.2 and 4.1).

- Guidelines for construction and development projects should be developed in such a way that a balance between building costs and green space implementation can be achieved. Promote the potential for space-saving construction and the unsealing of un- or underused impervious surfaces. In the case of infill development/re-densification, living quality has to be ensured (so-called 'double internal development' strategy in Germany). Guideline values for open space/green in the residential area and at the city scale (Sect. 3.7) should be binding (landscape plan, building legislation).
- Municipalities have to be equipped to develop urban green spaces according to their importance for the population and to maintain them through long-term care and management. Investment programs for nature and the environment in the city are required, as they commonly exist for housing and transport infrastructure. In existing funding programs, e.g., for flood or climate protection, 'green' solutions should be increasingly considered.
- As an instrument for the development of multifunctional urban areas, the intervention system (environmental impact assessment or compensation obligations) should be strengthened on the basis of current plans for land use and urban green spaces. Ecological compensation for newly built-up open spaces in the city area is needed (Sect. 4.1). Building greenery should be integrated into intervention and compensation schemes. Corresponding measures should be better controlled, e.g., via publicly accessible cadasters/presentations.
- Urban green space has to be recognized as cultural heritage and a tourism magnet (image of the city). Castles and gardens in Germany, temples, traditional residences, and garden facilities in China are part of the national cultural heritage, especially in urban regions, but they also provide many habitat services as well as regulative and cultural ecosystem services (Sect. 4.2). This is also of importance for the economic development of cities.

In general, China adds much more new urban green space in new districts or even in new cities in a rapidly growing urbanization. In Germany, established cities often do not grow anymore or only do so slowly, or they are even shrinking, which opens new perspectives for greening inside the existing urban patterns. On a larger scale, China formalized its conception of green cities in the Integrated Reform Plan for Promoting Ecological Progress issued in 2015 which promotes respecting, protecting, and benefitting from nature. Protecting the environment, especially in urban areas, is the fundamental policy pursued with high priority. This brings the two countries close together again in their ideas to develop urban areas as environmental living spaces for the majority of their people, who can eventually benefit from urbanization in balance with nature.

Through top-down policies and strategies the development of good and effective 'Green Cities' concepts will be a focal point in both countries. These concepts can be developed on different scales: (a) for entire cities, and here China is a forerunner worldwide with the concept of 'Eco-Cities', (b) in new and established urban districts, and (c) in the small bits and pieces of green which can be integrated into new and existing urban patterns. Not only more urban green spaces are needed, but also green spaces of better quality to meet the needs of the people. The urban ecosystem approach can be a tool to synergize municipal strategies for both urban green spaces and biodiversity, together with engagement of urban people and for their benefit.

For stakeholders and local decision-makers (Fig. 5.1) the following main challenges (cf. Sect. 2.1; Box 5.1) can be addressed in regard to ongoing urbanization, competition and demand for scarce areas and resources using an integrative, ecosystem-based approach in cities.

(I) **Ensure green space quantity (proportion): Integration of all forms of urban vegetation into a coherent green-blue infrastructure network strategy**

Recommendation/Action goal	Examples	To be found in book section…
Use modern concepts and binding, negotiated, political strategies; set policies, targets and incentive mechanisms to reduce urban land consumption; set own strategies and concepts for your city/district to address the local challenges, requirements, opportunities; work with standard/target values for urban green spaces in your city	– Ecosystem service concept, city biodiversity strategy and action plans – Less than 30 ha aim of the German government for maximum daily land take; 'no-net-loss' strategy of the EU; create incentives to avoid soil sealing – Eco-city/Garden-city concept defined by the Chinese government – Refer to benchmarks (e.g., WHO recommends at least 9 m^2 green spaces/inh.)	2.1/2.2 (strategies and concepts) 3.7 (standard/target values)
Ensure the optimum distribution of green space	– Compact city in a green network – Protection and low-impact development of urban green spaces, waters and wetlands by allowing natural development	4.3.6 (*Leitbild* of the city of Dresden)
Try to establish new green elements on redeveloped sites; link grey and green infrastructure	– Creation of new parks, gardens, and grasslands – Roadside greenery – Renaturation of brownfields – Unsealing	4.3.4 (Munich) 4.3.3 (Shanghai)

(continued)

(continued)

Recommendation/Action goal	Examples	To be found in book section…
	– Preservation and protecting/keeping a wide variety of small-scale green spaces (e.g., backyard greening, roof and wall greening, pocket parks in and near residential areas)	
Promote local self-sufficiency of basic resources to approach regenerative cities; strengthening all forms of urban agriculture; strengthening of roof and wall greening; improve the natural resource regeneration capacity of the city	– Protection of agricultural areas – Promotion of community gardens, allotments – Assess supply and demand of ecosystem services related to urban green spaces; promote measures	3.5 (providing ecosystem services) 4.3.2 (gardening in Berlin)

(II) Enhancement of ecological, social and/or economic qualities of urban green spaces (functionality, design)

Recommendation/Action goal	Examples	To be found in book section…
Promote biodiversity, strengthen the networking function of green infrastructure, emphasize the benefits of protected areas and species	– Habitat diversity: opportunities for species diversity, source of nutrition, breeding, and retreat habitats – Integration of natural succession into urban green systems/urban green infrastructure, allowing dynamics – Nature conservation areas – Overall city-wide biodiversity strategies	3.2, 4.2, 4.3 4.3.8 (Bonn)
Increase the stock of urban trees (adoption of tree protection statutes in all cities), secure and protect trees	– Protection of old/outstanding trees – Protection of trees during construction work – Tree planting, afforestation, tree preservation bye-laws	3.3.3 (trees help improve air quality) 4.3.1 (massive tree planting in Beijing)
Raise awareness of the ecosystem services provided by urban green spaces	Urban green spaces improve: – Health of residents, quality of life – Climate–water–air/energy conditions – Recreation opportunities – Reduce stress and anxiety	4.3.1 3.1, 3.3, 3.4 Box 4.1

(continued)

(continued)

Recommendation/Action goal	Examples	To be found in book section…
Upgrade existing areas; industrial restructuring linked to land renewal	Conversion/transformation of brownfields, former mining areas, etc.	4.3.6 (Xuzhou)
Create spaces for nature experience; avoid interventions in green areas, which serve for sports, play, and movement	Develop urban green spaces according to the demands of different population groups such as children and elderly	4.2, 4.3.8 (Bonn)

(III) **Use optimal planning approaches (treated in depth in Sect. 5.2), instruments, governance tools, etc.**

Recommendation/Action goal	Examples	To be found in book section…
Set up city profiles, strategic targets (labeling); collect data/information (creation, collection, access); capacity development (knowledge, organizational capacity)	– Shanghai will become greener – Wenjiang district in Chengdu as pilot district ('Garden city') – *Leitbild* Dresden (compact city in an ecological network)	2.3, all case studies in Sect. 4.3
Governance: laws, regulations, policies; institutions (e.g., transparent standards and costs)	– Compensation mechanisms, obligatory planning, partnerships, decision-making process, policy integration/coordination – Ecosystem services as guiding principle	4.1 4.3.1(Beijing)
Use 'planning principle of countervailing influence' (mixed planning or top-down/bottom-up planning)	– Existence and input of plans to guide the work of green space planners – Integration of green space planning into other kinds of planning or linkage between them	4.1
Enhance stakeholder participation and public involvement; education, identification, and communication, iterative process for knowledge production	– Include wishes of the people (demand) – Create incentives such as subsidies or competitions to promote greening by residents and investors	3.4, 4.2, 4.3 4.3.4 (Munich)

(continued)

(continued)

Recommendation/Action goal	Examples	To be found in book section…
Perform economic valuation of ecosystem services; cost–benefit analyses; find innovative financing models, (developing assistance and funding strategies, taxes, loans, lotteries); reduce maintenance and management costs	– Economic assessment has the potential to raise public awareness of the value of urban green and provides helpful information for sustainable urban development; – Greening benefits expressed in economic terms complement conventional ecological–environmental emphasis (incentives) – Local fiscal systems should redistribute parts of the urban value generated – Public–private partnerships for urban green/biodiversity	Chap. 3, in particular Sect. 3.6, 4.1.3 (The findings of various cases studies demonstrate the consequences of land-use changes and loss of urban green spaces for the underlying ecosystem services and provide decision-makers with additional information in the process of urban planning.)
Help to qualify social neighborhood development → environmental justice	Formulate and implement standards for urban areas (access for everyone…)	3.4, 3.7, 4.2

5.2 Guideline on How to Deal with Complex Objectives and Various Scales in the Course of Urban Green Space and Land-Use Planning

Martina Artmann, Olaf Bastian, Jiang Chang, Boping Chen, Karsten Grunewald, Tinghao Hu, Juliane Mathey, Stefanie Rößler and Wei Hou

Cities are shaped by complex compositions of different land cover and land-use types, unlike any other kind of landscape. In guiding cities toward sustainability, land-use planning is an important management instrument. Planners can act as mediators for balancing social, economic, and environmental interests arising in the city, in individual parts, or on different land-use types. However, in cities space is limited, and balancing objectives related to environment, economy, and society are challenging. In particular, urban land-use and green space planning need to deal with trade-offs between objectives related to further urban development, such as for residential or commercial purposes (resulting in an increase in soil sealing) on the one hand, and the protection and (re-)development of urban green spaces on the other hand.

Besides dealing with various objectives, urban land-use and green space planning should be organized following the idea of counterflow: Top-down planning, setting overall ambitious objectives and rules beginning from a national, state or regional perspective, needs to be accompanied by bottom-up planning from the site and local level to ensure the inclusion of local needs and knowledge for promoting appropriate and feasible objectives and measures. Higher policies such as the European Green Infrastructure Strategy (EC 2013) can set important signals and promote the implementation of ecosystems services by urban planning at the municipal level (Hansen et al. 2015). However, in Germany the main responsibility for tasks such as managing urban soil sealing is placed at the city scale, where urban planning has the main authority to develop sealing and green space planning strategies and to put them into practice (Artmann 2014). In China, setting central level objectives on urban land use, green space, and soil sealing can be a fitting tool, but policies and mechanisms to motivate the local government to effectively implement such objectives need to be developed.

At present, the green space development and planning system in China forms a top-down complete regulation framework from the national to the local level (Sect. 4.1.1). The plan for National Economic and Social Development (also called 'Five-Year Plan') acts as the top guidance and as a steering wheel, specifying aggregated indicators rather than specific measurements. 'Environmental protection and ecological control program' is a generic term for a series of plans that play an active role in promoting green space development. This category of plans further extends and refines the relevant provisions of green space development and environment protection requirements. The land-use plan is one of the most strict land management means, which can be implemented at the national, provincial, urban, and county levels. As a result, it directly decides the scale, function, and structure of green spaces in spatial aspects. Based on the land-use plan, the urban master and urban green system plan will further determine the layout and form of green spaces at the local level.

Local governments also organize the compilation of some informal plans that closely relate to the development and protection of urban green spaces. These plans are a powerful necessary supplement to the urban master and urban green system plans. Additionally, a series of urban gardening actions have been rapidly carried out in China. This not only helps to protect and develop urban green space and the environment, but also to promote awareness and appreciation of the city, as well as awareness of responsibility and of environmental protection among the citizens.

Within the Chinese context, the following issues still need to be further solved to better implement urban green space development.

(1) In general, the green space development and planning system in China is complicated and the durations of plans are relatively short. As a result, how to further promote an orientation toward science and enforcement still remains to be solved.
(2) Because of different goals and orientations of various strategies, trade-offs between ideas and aspects of green space development and green system

planning may exist in plans on the same level. Therefore, a ‘Multiple Planning Integration’ approach is central to planning system reform in China.

(3) The legal status and enforcement of local non-statutory plans remain to be further enhanced.

Within the German context, a number of instruments are available to deal with issues of ‘urban biodiversity’, ‘urban ecosystem services’, and ‘green space development’ (Sect. 4.1.2). To be successful:

(1) mainstreaming of these issues into spatial planning instruments is needed;
(2) individual issues need to be addressed at the appropriate scale;
(3) all planning levels, from regional through city-wide and district to site level, need to be considered;
(4) formal instruments cannot be replaced but must be supplemented by informal instruments.

In Germany, landscape planning is an important tool to shape urban land use and land cover and to set visions toward sustainable urban development (BfN 2008). Landscape planning includes at least a two-stage landscape planning process: the regional landscape plan is based on the landscape structure plan, and the local regional plan is reflected by the landscape plan. The landscape plan identifies nature conservation objectives and interlinked management guidelines on the local scale. Green infrastructure planning and ecosystem services can support landscape planning in structuring the complex task of sustainably managing urban ecosystems and their environment on different scales. The concepts of green infrastructure and ecosystem services can support for instance landscape planning in approaching compact and green cities as a crucial challenge arising in urban land-use planning (Sect. 4.3.6). Green infrastructure and ecosystem services can support German landscape planning in the future and provide opportunities for retaining challenges of compact and provide opportunities for retaining challenges of compact cities. The concepts of green infrastructure and ecosystem services support landscape planning in reflecting cities as socio-ecological systems (e.g., different demands by urban residents on the urban ecosystem such as recreation or contact with nature) and considering spatial heterogeneity and properties of urban ecosystems and landscapes (e.g., different types of urban green spaces such as urban parks or allotment gardens). The guideline for compact and green cities reflects three major modules in German landscape planning: (I) analysis and evaluation of landscapes; (II) planning targets and measures; and (III) the impact assessment of planning targets and measures. In each module, the vision of a compact and green city is considered by reflecting green structures and governance processes as part of green infrastructure planning. The multi-functionality of green infrastructure is reflected by the supply of various ecosystem services (Table 5.1). A multi-scale green space

Table 5.1 Considering green infrastructure and ecosystem services in landscape planning (LP) to build compact green cities, on the example of Germany (Artmann et al. 2017)

	(I) State of nature and landscape	(II) Planning targets and measures	(III) Planning impacts
(1) *Guidelines for considering green structures within landscape planning for compact green cities*			
(1.1) Multi-object approach	LP classifies different types of urban green space, e.g., urban parks, forests, and private gardens (see Table 5.2)	LP formulates planning targets and measures for different types of urban green space, e.g., targets for provision of private and public areas (see Table 5.2)	LP evaluates the impacts of planning targets and measures on different urban green space types, e.g., impacts on urban parks (see Table 5.2)
(1.2) Integration	Within LP the status of integration of urban green space in developed areas is analyzed, e.g., the provision of roadside trees in urban centers (see Table 5.2)	LP formulates planning targets and measures to integrate urban green spaces into developed areas, e.g., increase in *per capita* green spaces in the urban center (see Table 5.2)	LP evaluates the impacts of planning targets and measures to integrate urban green spaces into developed areas, e.g., risk of noisy recreational activities in parks (see Table 5.2)
(1.3) Connectivity	LP considers the connectivity of urban green spaces, e.g., the process of connecting green spaces between the city center and the urban fringe	LP formulates planning targets and measures to connect urban green spaces functionally and physically, e.g., connecting recreational spaces through roadside trees	LP analyzes the impacts of planning targets and measures for green infrastructure connectivity, e.g., impact of green networks on biodiversity
(1.4) Multi-functionality	The status of green spaces providing ecosystem functions, e.g., supply of fresh air in the city center, is analyzed within LP	LP formulates planning targets and measures for ecosystem functions provided by urban green spaces, e.g., increased cooling capacity in highly sealed districts	LP evaluates the impacts of green infrastructural measures on ecosystem functions, e.g., effect of open private green spaces on reducing heat stress

(continued)

Table 5.1 (continued)

	(I) State of nature and landscape	(II) Planning targets and measures	(III) Planning impacts
(2) *Guidelines for considering green governance processes within landscape planning for compact green cities*			
(2.1) Multi-scale approach	LP considers multi-scale regulations for compact green cities, e.g., the need to reduce land take according to national targets	LP formulates planning targets and measures in alignment with multi-scale regulations for compact and green cities, e.g., implementation of green networks under national nature conservation laws	LP evaluates impacts of planning measures by reviewing multi-scale targets for compact and green cities, e.g., national targets to foster the compact city
(2.2) Strategic approach	LP analyzes the state of green infrastructure in relation to urban sprawl	LP formulates planning targets and measures for reducing urban sprawl through green infrastructure	LP evaluates measures to reduce urban sprawl through green infrastructure
(2.3) Social inclusion	LP considers different actor groups for the evaluation of nature and landscape, e.g., impact of climate change on vulnerable population groups	LP formulates planning targets and measures for various actor groups to realize compact and green cities, e.g., by motivating residents to create green buildings	LP evaluates the impacts of planning targets and measures on various actors, e.g., impact of limiting urban sprawl on farmers
(2.4) Transdisciplinarity	LP uses expertise from various disciplines to analyze the status of the compact and green city, e.g., research on noise pollution	LP formulates planning targets and measures for compact and green cities using expertise from various disciplines, e.g., scientific models of climate regulation	LP evaluates the impacts and conflicts of planning measures using expertise of different disciplines for compact and green cities, e.g., scientific findings on the impact of green roofing on climate regulation

planning and integration of the green into the gray infrastructure is also considered in the guideline by applying the multi-object approach of green infrastructure planning (Table 5.2).

The guideline developed was tested on the example of the landscape plan for Dresden, which follows the *Leitbild* 'Dresden—the compact city in an ecological network' (Sect. 4.3.6). The results revealed that multi-scale urban green space planning and urban green space integration into the gray infrastructure were underrepresented in the landscape plan (Artmann et al. 2017). However, the land-use plan might be a more suitable planning instrument than the landscape plan to incorporate requirements on the greening of buildings at the site scale. Thus, the consideration of different possibilities of green space implementation needs to be addressed at the appropriate scale. However, so far in Germany not all potentials are used to implement the greening of roofs or walls, even though the greening of buildings can supply a range of ecosystem services, such as microclimate regulation, reduction of stormwater runoff, and support of biodiversity or food supply. Thus, further efforts in legislation, financial incentives, research, and education are necessary to foster the implementation of green buildings (Table 5.3). In general, urban planning should make use of a mix of instruments and strategies to foster green cities by reflecting various actors and spatial scales.

Table 5.2 Multi-object and multi-scale planning of green infrastructure in landscape planning (Artmann et al. 2017; based on Landscape Institute cited by EEA 2011; Davies et al. 2015)

LP considers integration of green infrastructure into the built environment: – Roadside trees and hedges – Green buildings (e.g., green roofs and facades) – Green space in built-up areas – Greenery in residential spaces – Greening of social infrastructure (e.g., schools) – Greening of commercial/industrial spaces – Greenery along transport infrastructure – Greening of water management systems – De-sealing/dismantling of built infrastructure	**LP considers urban green infrastructure at site scale**: – Pocket parks – Private gardens – Cemeteries – Ponds and streams – Small woodlands in developed areas – Playgrounds – Sports grounds – Greened city squares – Allotments – Vacant land
LP considers urban green infrastructure at city and district scale: – City/district parks – Forest parks – Lakes – Rivers and floodplains – Major recreational spaces – Brownfields – (Former) mineral extraction areas – Agricultural land – Viticulture	**LP considers urban green infrastructure at regional and national scale**: – Regional parks – Road and railway networks – Regional greenbelts – National parks – Open countryside – Long distance trails – Road and railway network

Table 5.3 Activities to promote green buildings in Germany (based on Fachverband Raumbegrünung und Hydrokultur 2016 with own adaptions)

Legislation and planning: – Green roofs and facades as a basic and obligatory requirement stipulated in legal regulations such as in urban development plans – Consideration of green buildings in the legal offset regulation – Initiation and promotion of urban green building strategies including an analysis of potential spaces where to green buildings and simulations for climate improvement	**Financial incentives**: – Direct monetary promotion of roof and facade greening provided by grants from the federal government and the states. Establishment of financial support programs – Provide grants for residents and investors to green their buildings additionally to legal requirements (Sect. 4.3.4) – Provide incentives through indirect promotion, such as rainwater charges and implementation of the split waste water fee in all major cities (Sect. 4.1.3)
Research: – Initiation and support of research projects. Establishment of a central research database – Transdisciplinary cooperation between research, architects, engineers, and planners (see also Artmann 2016) – Research is needed for: measuring evaporation performances, climate improvement, pollutant and fine dust binding, noise absorption, water retention during heavy rain, biodiversity, cost-benefit analysis, quality management, thermal insulation, and CO_2 binding	**Education**: – Workshops especially for cities, politicians and architects – Mandatory seminars on roof and facade greening for architects and city planners – Knowledge transfer through conferences, e.g., World Green Infrastructure Congress (WGIC) – Central website as a knowledge platform and data pool: http://www.gebaeudegruen.info/

5.3 Potentials for Cooperation and Outlook

Karsten Grunewald, Martina Artmann, Olaf Bastian, Jürgen Breuste, Boping Chen, Tinghao Hu, Dennis Kalisch, Lennart Kümper-Schlake, Xiushan Li, Juliane Mathey, Stefanie Rößler, Ralf-Uwe Syrbe, Wei Hou, Henry Wüstemann and Suili Xiao

Increasing urban populations and pressure for urban development will continue to challenge the provision and the quality of green spaces and green infrastructure now and in the future. On the one hand, the prospects for green space may depend on the ability to achieve convincing preservation. On the other hand, decisions about urban green development are made by society and urban dwellers. In this context, systematic thinking and a holistic approach are necessary (as provided by the ecosystem service concept, Grunewald and Bastian 2015).

Natural capital and ecosystem conditions/services will become more and more critical with population growth and increasing demand for natural resources and space. Additionally, we can expect further obstacles and challenges, including ongoing climate change, aging population, changing transport systems (e.g., electric cars), increasing energy consumption, and a further digitized world. International standards and strategies for the reduction of land consumption and the development of goals, indicators, and monitoring concepts are necessary.

The promotion of sustainable urbanization in previous years has led to many positive experiences. It is believed that sharing and learning best practices between different countries can make significant contributions to the global mission of sustainable urbanization (Shen et al. 2013). In order to share experiences of green city development, there is a need for properly evaluating the performance of the implemented approaches and identifying best practices.

In addition to the existing EU-China Urbanization Partnership, the Sino-German Urbanization Partnership has been initiated in May 2013 by the German Chancellor Angela Merkel and the Chinese Premier Li Keqiang. The German Federal Ministry for the Environment, Nature Conservation, Building and Nuclear Safety (BMUB) and the Chinese Ministry of Housing and Urban-Rural Development (MoHURD) are jointly responsible for implementing activities based on this partnership (BMUB 2016a). The aim is to develop a climate-friendly, integrated and sustainable urban development policy through political dialogue, city–city exchange of experience, and training with a practical emphasis. A Joint Declaration of Intent between BMUB and MoHURD was signed in November 2015 and 4.8 Mio. € until 2020 were allotted to the Deutsche Gesellschaft für Internationale Zusammenarbeit GmbH (GIZ) to implement the Sino-German Urbanisation Partnership in the framework of the International Climate Initiative of BMUB (Müller 2015).

The Sino-German Mayors' Program was launched in 1982 and continues to operate with one summit annually. The insights and experience gained from the cross-city dialogue events and the Mayors' Summit are being fed into the policy dialogue between Germany and China. On this basis, policy recommendations for decision-makers at the national, provincial, municipal, and local levels are being developed. The Sino-German Environment Forum, which takes place every three years and is organized by BMUB and the Chinese Ministry of Environmental Protection (MEP) together with the industry. It covers topics such as air pollution control, water management, resource and energy efficiency, the carbon market, biodiversity, green procurement, sustainable business, sustainable consumption, and environmental labeling (BMUB 2016a).

One focus of the Sino-German cooperation is in the field of biodiversity and ecosystem services, *inter alia* because this ensures connectivity to international processes in nature conservation policy such as MEA, CBD, TEEB, and IPBES. In China, air, water, and soil are also attracting the attention of its environmental policy. Starting from partially critical states of ecosystems, fundamental strategies will be developed and implemented in the near future in the area of conservation and restoration of ecosystems and their services as well as biodiversity, especially in urban and peri-urban areas. The authors of this book are especially involved in this process, and intensive exchange (e.g., joint workshops and real green-city-lab field studies in Berlin, Dresden, Chengdu, Shanghai, Beijing, Nanjing, Xuzhou, Xianju in 2015/2016) has been incorporated into this study.

In European cities, urban green space development has been a hot topic for more than 100 years. Green networks with parks and urban forests have been taken into account in the expansion of cities. For example, in Dresden, after a flood disaster in the middle of the nineteenth century, the town authorities recognized the value of the floodplain meadows along the Elbe River for the general public and kept them free from construction by law. However, over the past decades low-density suburban development in the periphery of Europe's cities has become the norm, and in many European countries the expansion of urban areas has exceeded the population growth by more than three times. On the one hand, traditional environmental health problems from unsafe drinking water, inadequate sanitation, poor housing, or air pollution are mostly solved or reduced. On the other hand, the sprawling nature of the cities in the Western urban world is critically important because of major impacts that are evident in increased energy, land, and soil consumption, which threaten natural, urban, and rural environments. The rising greenhouse gas emissions, which cause climate change and raise air and noise pollution levels, often exceed the agreed human safety limits (e.g., Uhel 2008).

Germany already started with research on urban ecology and with implementing the results in practice in the late 1960s (Sukopp and Wittig 1998); Twenty years later, China initiated its basic research in this field. 'Green city development and urban biodiversity and ecosystem services' became a subject in Chinese urban ecological research less than 10 years ago. Despite the heated international discussion on ecosystem services and their enhancement, so far administrative institutions responsible for the planning of German cities have been hesitant to take up the issue (e.g., Haase in Grunewald and Bastian 2015).

Current subjects of Chinese studies are spatiotemporal patterns and driving processes of urbanization, urban growth modeling, urban heat islands (many studies, Sect. 3.3.1), environmental impacts of urbanization, urban ecosystem services, sustainable cities/eco-cities, and urban sustainability assessment. The interaction of research and practice has only recently been developing in the field of urban ecological research and development. Besides universities and institutes, such as the Chinese Academy of Science, schools also play an important role in Chinese research on the subject.

Germany can learn how the implementation of greening ideas can be executed in well-managed large-scale top-down steered projects and German enterprises; landscape planners and architects can participate in this process. China can learn about technologies and facets of urban greening and include this knowledge and experience in its own developing projects. The political level expects results which (still) do not exist and perhaps need more time or different stimulations. To bridge this gap, policy declares 'positive results' which lack profound research (e.g., awarding of eco-cities by the Ministry of Civil Affairs on merely statistical criteria). This on the one hand frustrates the very motivated scientists in this field, and on the other hand indicates a need for more research on how to deliver effective results.

Potential starting points for a collaboration at the interface between science and politics in both China and Germany are as follows:

- In China, politics is a top-down subject broadly accepted as such in society. Only with political support can specific subjects be included in research to legitimate them. A subject in research must be recognized by political institutions before intensive research occurs.
- Urban ecological research has been recognized as something worth supporting from the highest political level in China. This allows getting political (and financial) support for such research. But the political level also affects the fields the research should focus on, since political programs are set from time to time and may indicate their acceptances of results. Research subjects and targets sometimes become political slogans (e.g., 'national park city', 'eco-city', 'eco-cultural development', etc.).
- The political level demands practically applicable results but has not explored clear ways to process them. That is why the cooperation between research and practice is still lacking.

Further topics of collaboration between China and Germany:

Both countries are addressing the challenge to approach compact cities by limiting urban sprawl. However, to provide a city of high living quality the compact city concept should integrate the aspects of the protection, qualification, and (re-)integration of urban green spaces. This concept is accepted by both sides, China and Germany, and it is addressed by their governmental and scientific institutions. The Sino-German collaboration in terms of knowledge and good practice exchange can support the process of making cities greener by recognizing the multiple benefits of urban green spaces that are often neglected in contemporary urban planning.

The concepts of green infrastructure and ecosystem services provide guidelines on how to interlink the 'urban gray' with the 'urban green'. Case studies in Germany and China can exhibit how both approaches are implemented in planning practice for compact and green cities. A main focus for future research on cities as socio-ecological systems can be the aspect of social inclusion as a major pillar of green infrastructure planning. How can urban residents be motivated to green their city, where do benefits arise through bottom-up approaches, and how can the government, local authorities and others support such activities? In this regard, research needs to focus more on the demand side of ecosystem services to reflect cities as socio-ecological systems. Thus, to be able to respond to current challenges connected with urbanization, such as climate change, biodiversity loss, limitation of resources, and population growth, there is a need to integrate humans, their habits, and attitudes into basic and applied research more deeply (McPhearson et al. 2016). In China, but also in Germany, public awareness of the benefits of urban ecosystems needs to be improved so as to develop better public perception of nature and facilitate the overall ecological conservation process.

Research needs in the field 'Green city development and urban biodiversity and ecosystem services':

Making the monetary and non-monetary value of urban green spaces visible through accounting of ecosystem services could be a chance to consider external costs connected with the degradation of ecosystems. A better understanding of cross-scale dependencies between different ecosystem services is needed to get a clearer picture of urban ecosystem service flows as well as options and challenges for their safeguarding depending on scales of responsibilities and policy actions. A further question is how to manage urban growth processes such that negative socioeconomic, human, and environmental impacts of urbanization are minimized or avoided, and that socially integrative cities can develop in an environmentally friendly and financially viable way in order to provide favorable living conditions for the population.

Status of research in China with focus on urban green space issues:

- There is a lack of comprehensive Chinese studies of biodiversity and ecosystem processes in cities. The majority of researchers working on urban ecological subjects are geographers and environmental scientists who are skilled in remote sensing, GIS, and related technical areas, but lack training in fundamentals of ecology (Wu 2014).
- The Chinese Ecosystem Research Network (CERN; http://www.cern.ac.cn/) was established in 1988 by the Chinese Academy of Sciences. It contains 42 sites; only one revolves around urban ecology (in Beijing).
- There is a lack of tested conceptual frameworks systematically developed for urban ecological research in Chinese cities. Many studies are dealing with one or few aspects of urban ecological systems without considering the entire urban ecosystem.
- Considering the efficient top-down approach in China, assessment tools to evaluate urban green can be good policy tools. However, there is a need for further research on systematic, qualitative, and process-oriented assessment tools in order to ensure quality development of green cities rather than short-term political achievements.
- Most use-inspired or problem-oriented studies lack either ecological science or design practicality. The dual phenomenon of 'developing cities without planning' and 'planning cities without ecology' reflects the poor communication and collaboration between the two camps (Wu 2014).
- A novel, deep, and innovative/smart understanding of urban ecosystems is necessary, including landscape sciences, land system sciences, urban ecology and architecture, the relationship of urbanization to climate change and human health, the relationship between urban ecosystem services and human well-being, balancing environmental health, human welfare, and social equity in urban areas, and the relationship of biodiversity and quality of life in cities.
- A transdisciplinary science of urban regions is required to meet the national needs (Wu 2014).

- There is a tremendous lack of cooperation between different sciences and schools and different institutions which compete for research funding.
- Chinese urban researchers (ecologists, geographers, etc.), planners, and designers need to increase their collaboration.
- There is lack of unification of multiple regulations. In China, cities have separate plans, such as a plan for animal and plant diversity protection, a plan for urban biological diversity (urban resource and urban property), urban green space regulation, etc. They are also under the remit of different ministries which further complicates a unified approach.

A selection of recent research topics with focus on urban green space aspects in Germany (BMUB 2016b, 2017a), which topics can also help address some common urban challenges faced by both China and Germany (such as urban sprawl, soil sealing):

- Simulation projects to test new solutions for the green infrastructure and to develop concrete action strategies and instruments;
- Further development and testing of measures and instruments to reduce the amount of new soil sealing—compact city development, inner development, land recycling;
- Assessment/planning of climate protection and climate adaptation in the settlement sector in connection with ecosystem services/biodiversity;
- Assessment/planning of ecological impact of integrated inner-city development concepts on the infrastructures of urban expansion areas;
- Integration of ecosystems and ecosystem services into the environmental economic accounting—theoretical framework conditions and methodological principles;
- Planning-oriented steering/management of the development of urban settlements in urban/suburban regions, with particular emphasis on nature-adapted land-use and space-saving spatial planning;
- Investigation of consumer behavior and relevant consumer goods with regard to their impact on biodiversity and ecosystems worldwide with the derivation of recommendations for action and instruments for sustainable consumption as a contribution to the achievement of the Sustainable Development Goals;
- Investigations to gain knowledge for improving the implementation process.

We want to conclude with an outlook statement by Uhel (2008): "Reliable scientific arguments exist for shaping urban sustainability around the spine of environment quality and ecosystem services rather than simply based on meeting energy and transport demands. To this effect, we are more and more often confronted with long-term problems for which the outcomes are highly uncertain. Making sense in a complex world requires that we separate straightforward problems that can be solved through exchange of best practices, complicated ones where good practice helps, complex problems where practices are emerging and problems borne out of chaotic systems where novel practices are needed. If we want to seriously address the sustainability of our consumption and production we need to recognize uncertainties about the future, go beyond the short timescales of current policies and change our current preoccupation with working on many separate issues. We need to develop policies that reflect the complexity of the systems we are dealing with, so that we can address the needs of today's disenfranchised as well as those of future generations."

References

Artmann M (2014) Assessment of soil sealing management responses, strategies and targets towards ecologically sustainable urban land use management. Ambio 43(4):530–541

Artmann M (2016) Urban gray vs. urban green vs. soil protection—development of a systemic solution to soil sealing management on the example of Germany.Environ Impact Assess Rev 59:27–42

Artmann M, Bastian O, Grunewald K (2017) Using the concepts of green infrastructure and ecosystem services to specify Leitbilder for compact and green cities—the example of the landscape plan of Dresden (Germany). Sustainability 9(2), doi:10.3390/su9020198

BfN (2008) Landscape Planning. The basis of sustainable landscape development. federal agency for nature conservation (BfN).http://www.bfn.de/fileadmin/MDB/documents/themen/landschaftsplanung/landscape_planning_basis.pdf Accessed 1 Nov 2016

BMUB—Bundesministerium für Umwelt, Naturschutz, Bau und Reaktorsicherheit (2015) Grünbuch Stadtgrün. Grün in der Stadt—Für eine lebenswerte Zukunft. Bundesministerium für Umwelt, Naturschutz, Bau und Reaktorsicherheit

BMUB—Bundesministerium für Umwelt, Naturschutz, Bau und Reaktorsicherheit (2016a) Bilaterale Kooperation zu Umwelt- und Klimaschutz sowie nachhaltiger Urbanisierung mit China. http://www.bmub.bund.de/themen/europa-international/int-umweltpolitik/bilaterale-zusammenarbeit/bilaterale-kooperation-mit-china/. Accessed 14 Feb 2017

BMUB—Bundesministerium für Umwelt, Naturschutz, Bau und Reaktorsicherheit (2016b) Ressortforschungsplan des Bundesministeriums für Umwelt, Naturschutz, Bau und Reaktorsicherheit. http://www.bmub.bund.de/fileadmin/Daten_BMU/Download_PDF/Forschung/ressortforschungsplan_2016_gesamt_bf.pdf. Accessed 14 Feb 2017

BMUB—Bundesministerium für Umwelt, Naturschutz, Bau und Reaktorsicherheit (2017a) Ressortforschungsplan des Bundesministeriums für Umwelt, Naturschutz, Bau und Reaktorsicherheit. http://www.bmub.bund.de/fileadmin/Daten_BMU/Download_PDF/Forschung/ressortforschungsplan_gesamt_2017_bf.pdf. Accessed 14 Feb 2017

BMUB—Bundesministerium für Umwelt, Naturschutz, Bau und Reaktorsicherheit (2017b) Weißbuch Stadtgrün. Grün in der Stadt—Für eine lebenswerte Zukunft. Bundesministerium für Umwelt, Naturschutz, Bau und Reaktorsicherheit

Davies C, Hansen R, Rall E, Pauleit S, Lafortezza R, De Bellis Y, Santos A, Tosics I (2015) Green infrastructure planning and implementation. The status of European green space planning and implementation based on an analysis of selected European city-regions. http://greensurge.eu/working-packages/wp5/files/Green_Infrastructure_Planning_and_Implementation.pdf. Accessed 24 Mar 2016

EC—European Commission (2013) Communication from the Commission to the European Parliament, the Council, the European Economic and Social Committee and the Committee of the Regions. Green Infrastructure (GI)—Enhancing Europe's Natural Capital. COM (2013) 249 Final, EC, Brussels, Belgium

EEA—European Environment Agency (2011) Green infrastructure and territorial cohesion. The concept of green infrastructure and its integration into policies using monitoring systems. Publications Office of the European Union, Luxembourg

Fachverband Raumbegrünung und Hydrokultur (2016) Zum Weißbuch „Stadtgrün". Positionspapier zur Gebäudebegrünung. http://www.g-net.de/files/download/Position_ZVG_Gebaeudegruen_130716.pdf. Accessed 30 Jan 2017

Grunewald K, Bastian O (eds) (2015) Ecosystem services—concept, methods and case studies. Springer, Berlin

Hansen R, Frantzeskaki N, McPhearson T, Rall E, Kabisch N, Kaczorowska A, Kain J-H, Artmann M, Pauleit S (2015) The uptake of the ecosystem services concept in planning discourses of European and American cities. Ecosys Serv 12:228–246

McPhearson T, Pickett STA, Grimm NB, Niemelä J, Alberti M, Elmqvist T, Weber C, Haase D, Breuste J, Qureshi S (2016) Advancing urban ecology toward a science of cities. Bioscience 66:198–212

Müller S (2015) Towards the Sino-German urbanisation partnership. Presentation at the 8th Sino-German workshop on biodiversity conservation, Berlin, 18–19 June 2015

Qian J, Peng Y, Luo C, Wu, C, Du Q (2016) Urban land expansion and sustainable land use policy in Shenzhen: a case study of China's rapid urbanization. Sustainability 8(16). doi:10.3390/su8010016

Shen L-Y, Ochoa JJ, Zhang X, Yi P (2013) Experience mining for decision making on implementing sustainable urbanization—an innovative approach. Autom Constr 29:40–49

Sukopp H, Wittig R (eds) (1998) Stadtökologie, 2nd edn. Gustav Fischer, Stuttgart

Uhel R (2008) Urbanisation in Europe: limits to spatial growth. Keynote speech by Mr Ronan Uhel, Head of Spatial Analysis, European Environment Agency, at the 44th International Planning Congress, Dalian, China, 20 Sept 2008. http://www.eea.europa.eu/media/speeches/urbanisation-in-europe-limits-to-spatial-growth

Wu J (2014) Urban ecology and sustainability: the state-of-the–science and future directions. Landsc Urb Plan 125:209–221

The manufacturer's authorised representative in the EU is Springer Nature Customer Service Centre GmbH, Europaplatz 3, 69115 Heidelberg, Germany. If you have any concerns regarding our products, please contact ProductSafety@springernature.com

Printed and bound by CPI Group (UK) Ltd, Croydon, CR0 4YY
05/10/2026
02228122-0004